A GRAPHIC GUIDE
to the
MOLECULE that
SHOOK the WORLD

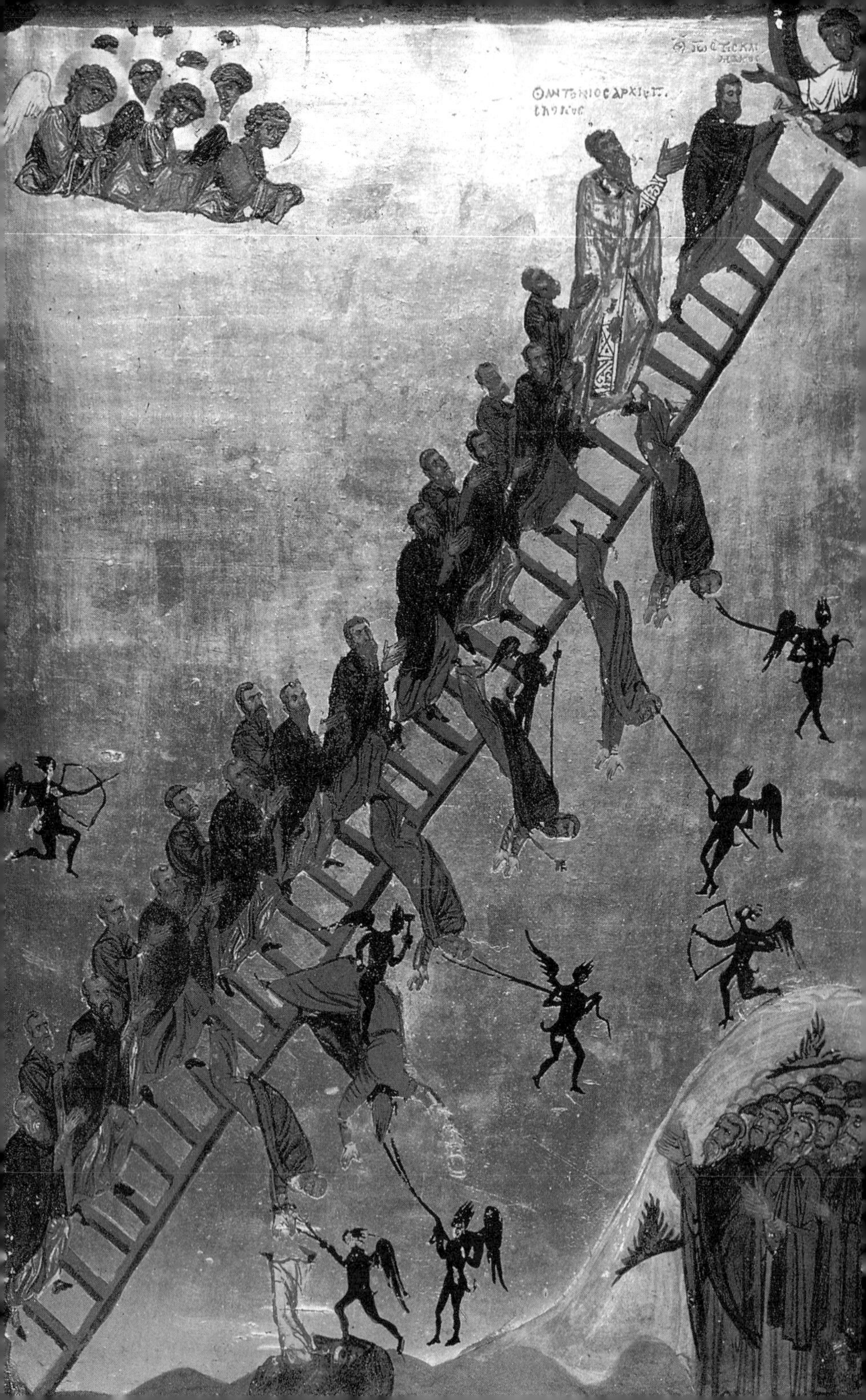
ΟΑΝΤѠΝΙΟϹΑΡΧΙΕΠ.

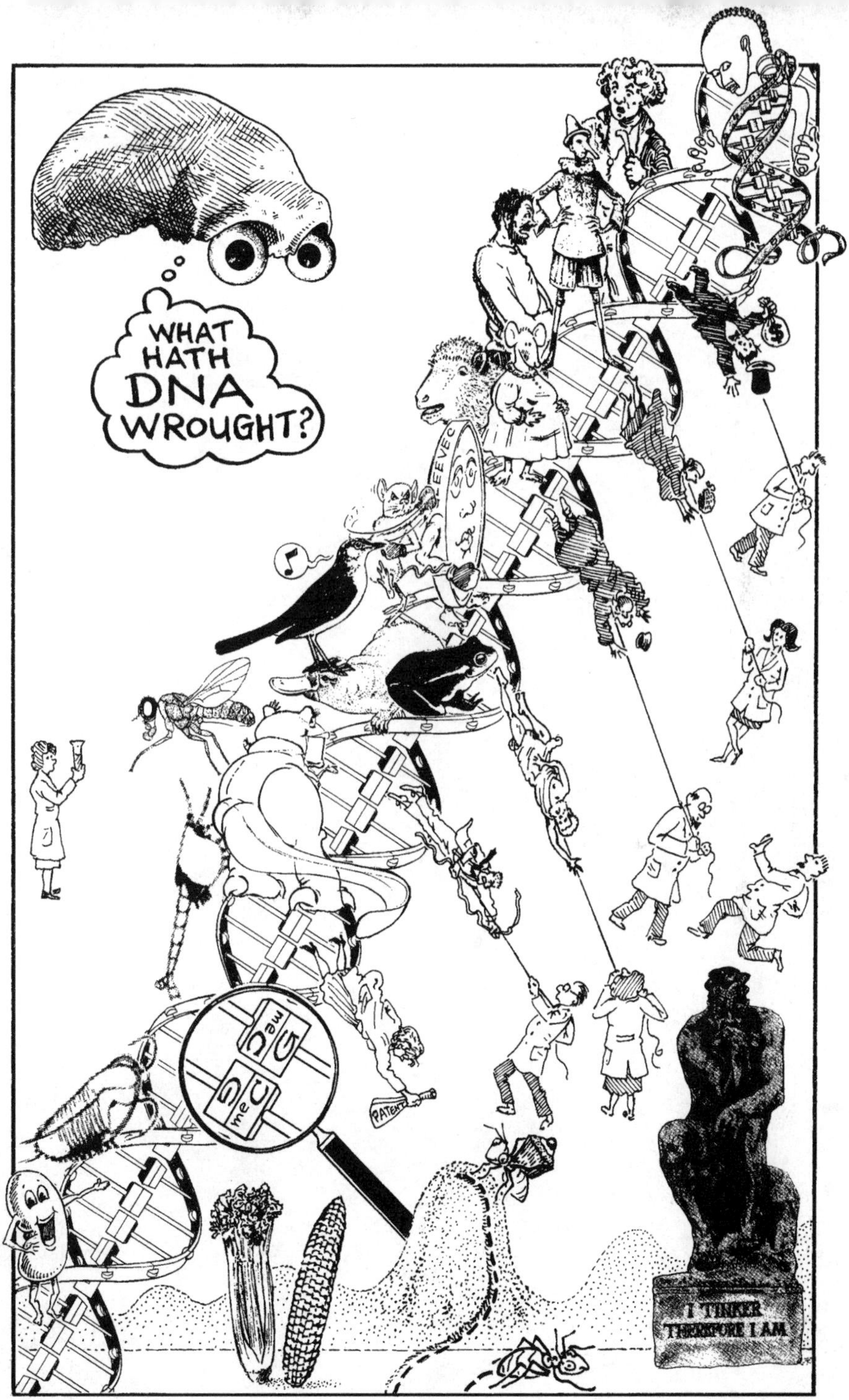
WHAT HATH DNA WROUGHT?
PATENT
I TINKER THEREFORE I AM

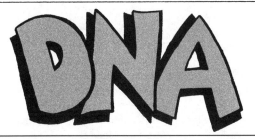

A GRAPHIC GUIDE to the
MOLECULE that SHOOK the WORLD

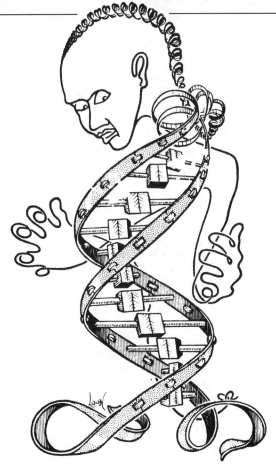

Israel Rosenfield, Edward Ziff, & Borin Van Loon

COLUMBIA UNIVERSITY PRESS NEW YORK

Columbia University Press
Publishers Since 1893
New York   Chichester, West Sussex

Copyright © 2011 Columbia University Press
All rights reserved

Library of Congress Cataloging-in-Publication Data
Rosenfield, Israel, 1939–
DNA : a graphic guide to the molecule that shook the world / Israel Rosenfield,
Edward Ziff, and Borin Van Loon.
p. cm.
Includes bibliographical references.
ISBN 978-0-231-14270-0 (cloth : alk. paper)
ISBN 978-0-231-14271-7 (pbk. : alk. paper)
ISBN 978-0-231-51231-2 (e-book)
1. DNA—Popular works. 2. Graphic novels. I. Ziff, Edward. II. Van Loon, Borin, 1951–
III. Title.

QP624.R67 2011
572.8′6—dc22

2010020889

Columbia University Press books are printed on permanent and
durable acid-free paper.
This book is printed on paper with recycled content.
Printed in the United States of America

c 10 9 8 7 6 5 4 3 2 1
p 10 9 8 7 6 5 4 3 2 1

Some of this material has appeared in a different form
in the *New York Review of Books*.
Image on p. ii: *Ladder of Ascent* / Art Resource.

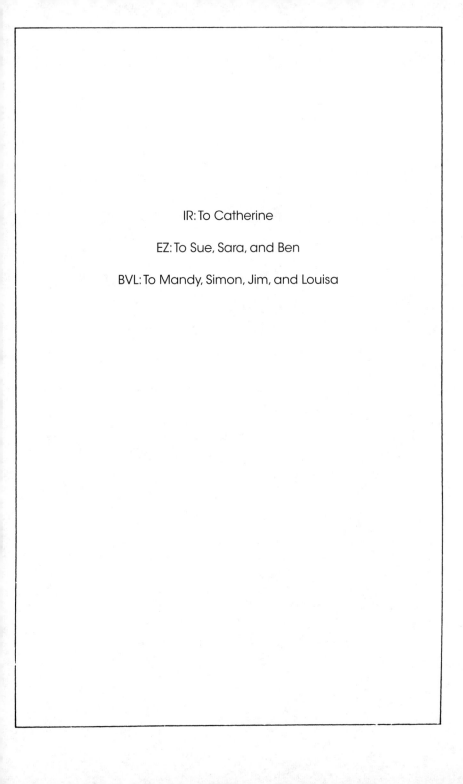

IR: To Catherine

EZ: To Sue, Sara, and Ben

BVL: To Mandy, Simon, Jim, and Louisa

Contents

Preface

Israel Rosenfield, Edward Ziff, & Borin Van Loon

This book is about DNA, the molecule that has a broad impact on our daily lives, affecting virtually all our activities. The original version of this book, *DNA for Beginners*, appeared at a time when DNA was already considered a fascinating subject, but no one would have foreseen that more than twenty years later its impact would reach almost every nook and cranny of human activity. The subject itself has radically changed. Our aim is to acquaint readers with this change and point to the deep moral, political, legal, financial, and scientific implications that now come into play.

The unveiling of the human genome has given us a new understanding of where we come from, who we are, and what we might become. It has shed light on our health and states of mind and affected our political and social relations. It has changed the study and practice of medicine – in terms of both diagnosis and treatment. The use (and abuse) of DNA in the courtroom has become common practice and has altered our notions of law. DNA has transformed our very understanding of what life is all about and has changed our view of evolution; we now have new insights into how our bodies are formed and a new understanding of our relation to other species.

It is our hope that this book combines humor, scientific depth, and philosophical and historical insights and that it will interest a wide range of readers, including those without any scientific or philosophical background, who seek to gain a sophisticated sense of the subject. The book can be read at several levels. Those not wanting to spend time with too many details can skim over or skip some of the more technical passages and can be easily carried along by the humorous illustrations, which often convey scientific and philosophical information. Some readers and students may want a deeper sense of the subject, and they may reread or spend more time on passages that delve into aspects of DNA that make it scientifically unique. It is important to realize that the humor and illustrations are as much a part of our conception of the book as the scientific text.

The idea for this revised version of the book came from Jim Jordan, the director of Columbia University Press. Its unusual format – the cartoons and the need for technical accuracy – required considerable assistance from the staff of the press, and we owe a special debt of gratitude to Milenda Lee for her gracious acceptance of our numerous revisions of both text and drawings. Our thanks to Patrick Fitzgerald for his editorial comments and to the two anonymous reviewers for carefully correcting and improving the text and drawings. We also thank Irene Pavitt for her fine proofreading of the book.

A GRAPHIC GUIDE
to the
MOLECULE that
SHOOK the WORLD

Introduction

DNA... three capital letters of which perhaps many may have heard, but not so many will have understood. This book is about the discovery of the importance of DNA, beginning in the mid-nineteenth century. We take a close look at the mechanisms believed to be important in its functioning, recognizing that much remains shrouded in mystery. And we examine the impact of DNA research on society, and some of the most recent findings.

Most DNA is in the form of a helical molecule. Some DNAs are very long, and some are small, but all DNA is so tiny that huge amounts of it can be found in the cells of living things. DNA holds the codes for an enormous variety of genes. And genes are the pieces of information which we all have inside us, enabling us to function and reproduce.

To get some idea of the great significance of DNA in the history and future of life on this planet, read on...

In December of 1949, almost four years before James Watson and Francis Crick published their model of DNA, launching the revolution in modern biology, the mathematician and designer of the computer, John von Neumann, gave a lecture explaining how a machine could reproduce. All it needs, he said, is a description of itself.

A machine with a magnetic core could not reproduce the magnetic core by making a mold. However, if it had a

description reading, "magnetic core: electric wire tightly wound around metal bar five hundred times, etc." and it had the necessary raw materials, it could easily follow the description and build the magnetic core.

The machine's offspring could reproduce as well, if the machine made a **copy** of the **description of itself** and inserted that copy into every new machine. Given the necessary raw materials, the machines could go on making copies of themselves.

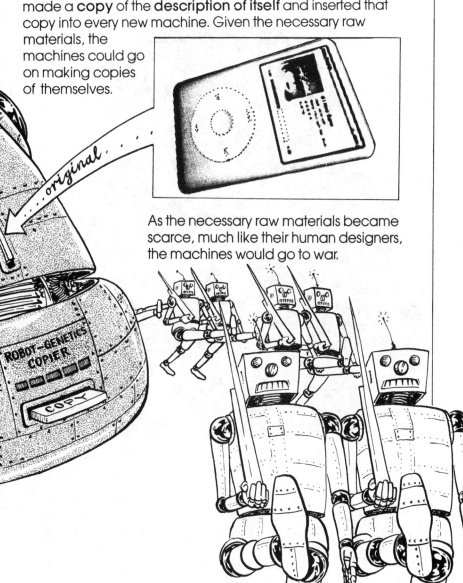

As the necessary raw materials became scarce, much like their human designers, the machines would go to war.

The **description** in von Neumann's machine is analogous to the DNA found in living things. Like the description of the machine, DNA contains the coded description of the organism and is responsible for its capacity to reproduce.

Living things, unlike von Neumann's machine, do not usually make **exact** copies of themselves. For then, there would be no evolution and life as we know it would not exist. Living things make **variant** copies of their parent organism or organisms. They can do this because of DNA.

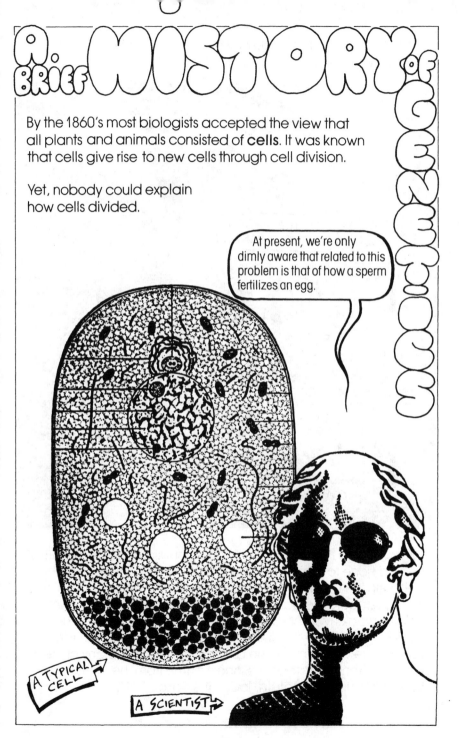

A BRIEF HISTORY OF GENETICS

By the 1860's most biologists accepted the view that all plants and animals consisted of **cells**. It was known that cells give rise to new cells through cell division.

Yet, nobody could explain how cells divided.

At present, we're only dimly aware that related to this problem is that of how a sperm fertilizes an egg.

A TYPICAL CELL

A SCIENTIST

DNA was discovered in 1869 by Frederick Miescher, who was then 25 years old.

Miescher was the son of a well-known physician in Basel. In 1869 he had gone to Tubingen to study the chemistry of white blood cells with the biochemist F. Hoppe-Seyler. He used pus obtained from postoperative bandages, as a source of the cells. When he added weak hydrochloric acid to the pus he obtained pure nuclei. If he added alkali and then acid to the nuclei a gray precipitate was formed. The precipitate was unlike any of the known organic substances. Since it came from the nucleus, Miescher called it **nuclein**. Today it is called DNA.

Shortly after Miescher's discovery, new staining techniques were developed which revealed band-like structures in the dividing cell that stained very darkly. In 1879 Walther Flemming introduced the term **chromatin** (*chroma*: Greek for "color") to describe the intensely stained material in the nucleus. In 1881 E. Zacharia found that chromatin reacted to acid and alkali in the same way as Miescher's nuclein. He concluded that nuclein and chromatin were one and the same.

TUBINGEN CASTLE
BIOCHEMICAL LAB.

The chromatin material observed in the 1880's was called chromosomes, the carriers of genes that are the basis of heredity. What is most remarkable is that some scientists studying fertilization made the connection between chromatin (chromosomes) and heredity already in the 1870's. Using the light microscope, Hermann Fol in Switzerland and Oskar Hertwig in Berlin independently observed that the sperm **penetrates** the egg and that the **nuclei** of the sperm and the egg **fuse**. And Edouard Van Beneden, studying the threadworm *Ascaris* (a parasite of horses) noted that the sperm contributed the same number of chromosomes as the egg to the developing embryo. He also discovered **meiosis**, the halving of the number of chromosomes in the germ cells (the egg and the sperm). It was Flemming who observed cells dividing and saw chromosomes replicating. He concluded that chromosomes were a source of continuity from one generation to the next.

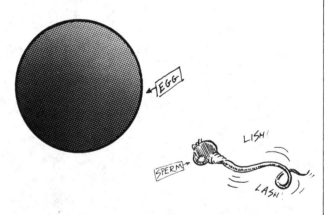

So by the 1890's scientists had come to have a clear idea of the nature of fertilization, and were even declaring that DNA (Miescher's nuclein) was the basis of heredity.

Modern genetics begins with Gregor Mendel's famous experiments with garden peas in the 1860's. Mendel had chosen peas that had certain pure traits which always breed true. He had plants that produced yellow seeds, and others that produced green seeds. When he cross-bred these types, all the progeny were yellow-seed-bearing plants. Mendel called the yellow trait DOMINANT, and the green RECESSIVE. He argued that the progeny of these first-generation crosses had each received an equal genetic contribution from each parent, but only the dominant yellow trait was manifested.

When he crossed these first-generation hybrids with each other he found that 75% of the progeny were yellow and 25% green, confirming his supposition that the green "gene" (he didn't use this term) had been there all the time. Mendel concluded that green and yellow were discrete genetic units which segregated independently into the progeny according to the laws of chance.

If chance governed the laws of inheritance of genetic information, where did the information needed to form a complicated biological organism reside? One answer, the so-called Vitalist School of Thought, argued that living organisms were shaped from outside by the hand of God. The Vitalists believed in EPIGENESIS, that the embryo developed from a simple unformed egg, gradually becoming a complex organism.

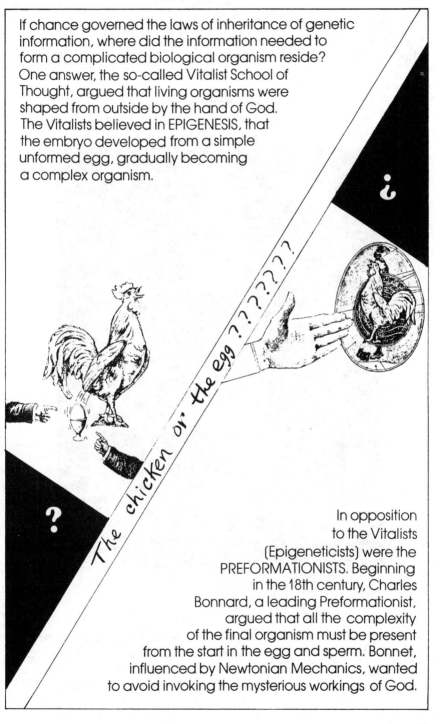

The chicken or the egg ???????

In opposition to the Vitalists (Epigeneticists) were the PREFORMATIONISTS. Beginning in the 18th century, Charles Bonnard, a leading Preformationist, argued that all the complexity of the final organism must be present from the start in the egg and sperm. Bonnet, influenced by Newtonian Mechanics, wanted to avoid invoking the mysterious workings of God.

11

By the 19th century, biologists could see the embryos develop under the microscope. Ernst Haeckel summarized another hypothesis with a familiar phrase ...

Ontogeny recapitulates phylogeny.

.... This means the early development of the embryo seems to repeat the **adult** stages of lower life forms from which it has descended. For example, at an early stage in its development, the human embryo has gills like a fish. As stated, Haeckel's law is no longer accepted.

Nineteenth-century scientists also wondered about the effect of environment on development. Mountain people have more red blood cells than their relatives living at sea level. And some people grow taller than others because of their diet. In 1809, Lamarck argued, among other things, that:

Changes brought about in an organism because of environment can be inherited. A tree that has been bent by the wind will give rise to bent trees.

This argument is usually summarized by the phrase: "acquired characteristics are inherited."

In Darwinian theory, natural selection – environmental pressures that affect an organism's chances of survival – operates on the innumerable variations among organisms. The variations are random, unlike the Larmarckian bending of the tree, whose character is directly imposed by the environment.

WITHOUT VARIATION ALL INDIVIDUALS WOULD BE IDENTICAL, AND THERE WOULD BE NO EVOLUTION.

The most obvious mechanism for creating genetic variation is sex. As with Mendel's peas, genetic traits are randomly segregated during sexual reproduction. The combinations, in a complex organism, are almost limitless.

When Darwin published **The Origin of Species**, the mechanism of fertilization of an egg by a sperm was not well understood. Darwin's explanation of fertilization (and the consequent variations) was derived from an old Greek theory: Pangenesis. Every organ and tissue secreted granules, called gemmules, which combined to make up the sex cells.

But Darwin's cousin Francis Galton transfused blood from rabbits of one color to those of another. The transfusions had no effect on the color of the offspring. Galton argued:

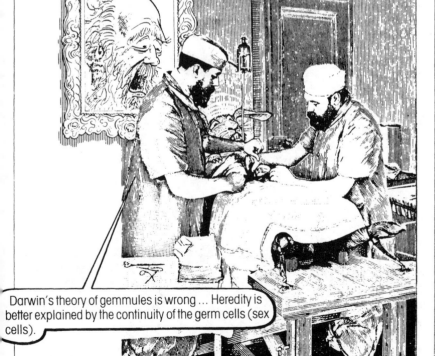

Darwin's theory of gemmules is wrong ... Heredity is better explained by the continuity of the germ cells (sex cells).

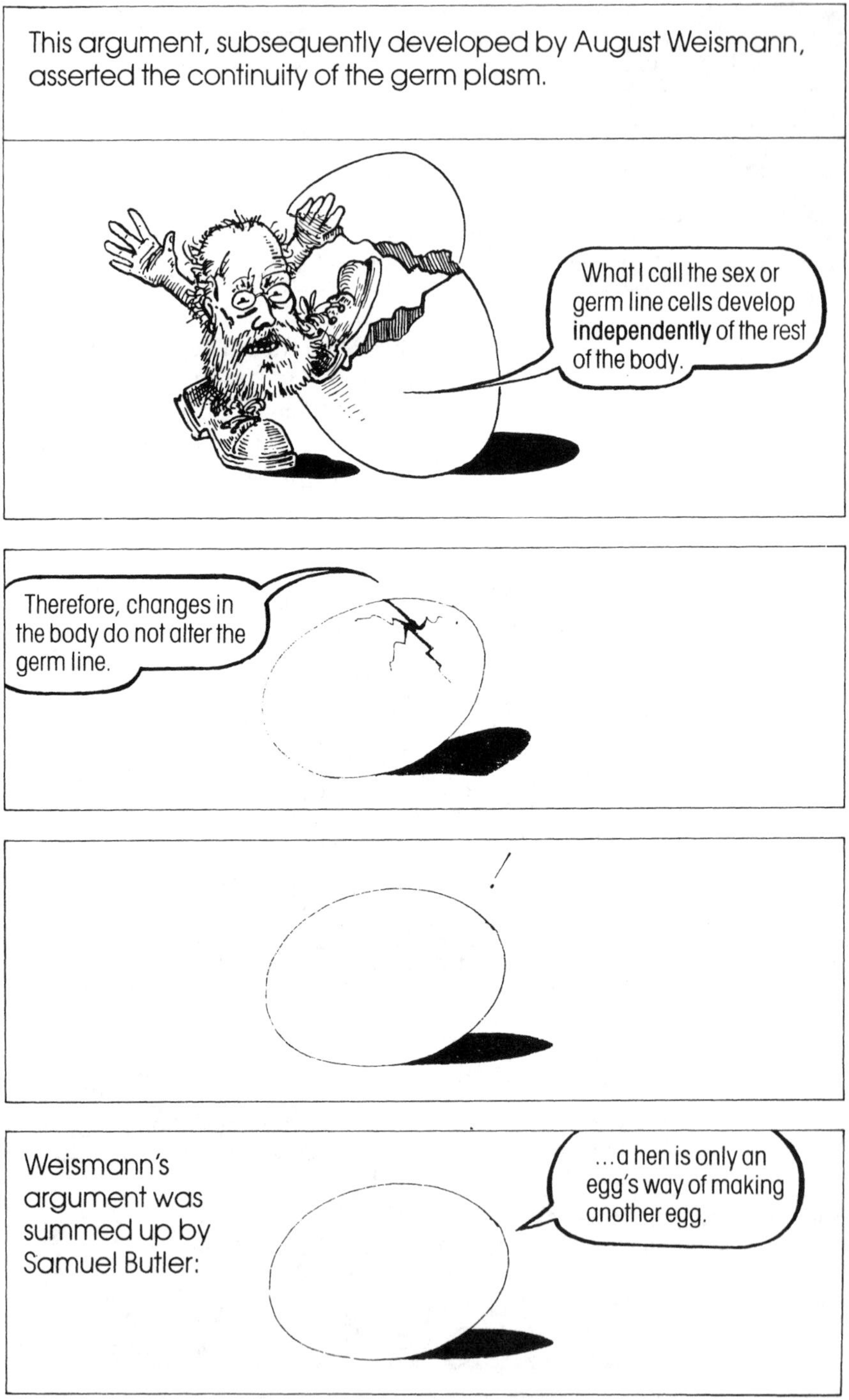
This argument, subsequently developed by August Weismann, asserted the continuity of the germ plasm.
What I call the sex or germ line cells develop **independently** of the rest of the body.
Therefore, changes in the body do not alter the germ line.
Weismann's argument was summed up by Samuel Butler:
...a hen is only an egg's way of making another egg.

Modern terminology describes this new understanding:

1) THE GENE:
   at the time considered a hypothetical unit that was responsible for carrying genetic information from one generation to the next.

2) PHENOTYPE:
   the observed properties of a living thing; a consequence of the interaction of genetic makeup with the environment.

3) GENOTYPE:
   the genetic makeup of the organism, as opposed to its physical appearance.

Let us recapitulate what beliefs were held by the early decades of the 20th century:

1) Mendel showed that genetic traits were discrete units which assorted independently.

2) To explain development, Bonnet and the Preformationists had argued that complexity must be present in the egg and the sperm. The epigeneticists argued that the embryo developed from a simple to a complex form shaped by external forces.

3) Haeckel suggested that ontogeny recapitulates phylogeny.

4) Darwin's theory of evolution is based on the notion of **variations** (produced in part through sexual reproduction) that are selected by environmental forces **(natural selection)**.

5) Lamarck asserted inheritance of acquired characteristics.

A controversy over the mechanisms of evolution and development was raging. It could not begin to be resolved until scientists understood DNA.

At the turn of the century genetics was a young discipline. The early geneticist was little more than a statistician of inheritance. To speed experiments, geneticists turned to the fruit fly DROSOPHILA. The DROSOPHILA life cycle takes about 14 days, and the flies are easy to breed, cheap to grow, and genetically rather simple because they contain only four chromosome pairs.

The early Drosophila geneticist Thomas Hunt Morgan observed violations of Mendel's Law of independent assortment of genes. Certain genes remained linked together in crosses more frequently than predicted by Mendel's statistics.

Morgan found four groups of linked genes in Drosophila. The tendency of genes to remain together in offspring suggested that they shared a physical association, and were joined together on the same chromosome. Drosophila had four "linkage groups" because it had four chromosomes.

Morgan next observed a low frequency of "assortment" even for traits that were on the **same** chromosome. Morgan suspected that chromosomes could break and recombine, allowing genes on the same chromosome to reassort. Genes far apart on a chromosome would have a greater chance of a break occurring between them than genes situated close together. If so, reassortment frequency would be a measure of gene distance. Using this prediction to test his break and rejoin model, Morgan was able to make maps of genes on the *Drosophila* chromosomes. Morgan's important finding was that genes fall in a defined linear order, and occupy specific positions on chromosomes.

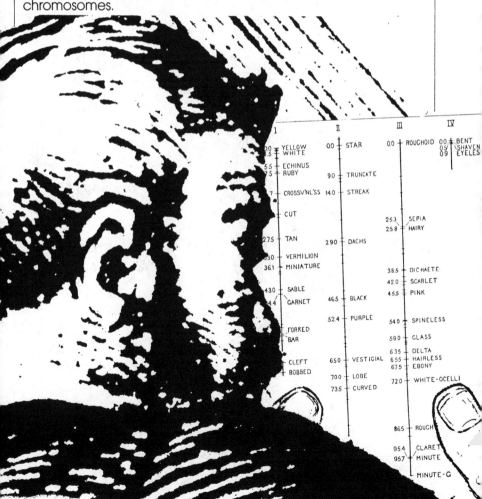

In the 1930's, scientists had little hope of "grinding a gene in a mortar, or distilling it in a retort." However, one property of a gene which could be analyzed was its ability to **mutate**. Hans J. Muller, a student of Morgan, increased the mutation rate of *Drosophila* 15,000-fold over the natural rate by X-ray irradiation of the flies. Muller appreciated that mutations resulted from chemical reactions, or "sub-microscopic accidents" produced by the X-ray beam in the genetic material. Mutations never observed in nature, such as "splotched wing" and "sex-combless," as well as natural ones including "white eye," "miniature wing," and "forked bristles" were found.

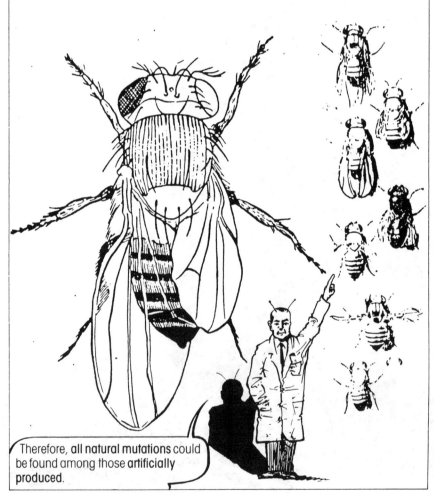

Therefore, **all natural mutations** could be found among those **artificially produced.**

Muller moved to Russia from the United States, where he disputed with the Soviet geneticist T.D. Lysenko.

I take the Lamarckian view that environmentally acquired characteristics are inherited.

No! After all, "white eye" and other natural mutations appeared in the "unnaturally" X-ray mutagenized flies. Lysenko's "environment" wasn't required to produce the mutations. Only the chemical changes produced by the X-rays were necessary. Mutations, either natural or lab-induced, provide the raw material for Darwinian selection.

Muller's experiments proved Lysenko erroneous.

Genetic damage can result from modern uses of radiation. I would also advocate eugenics, the regulation of reproductive behavior to favor special physical, intellectual, and moral qualities.

Although the dispute with Lysenko drove Muller from Russia he continued to speak out on the factors which shape the genetic constitution of society.

Puzzlement remained about how a gene actually operated. It would soon be shown, first by Sir Archibald Garrod in London, and later by the Americans George Beadle and Edward L. Tatum, that a gene specified an enzyme.

An enzyme is a type of protein that catalyzes biological reactions. By specifying an enzyme, a gene enables a particular chemical reaction to take place in a cell.

I summarize this in my "one gene–one enzyme" hypothesis.

Ah, but now we must identify the chemical substances in chromosomes which store and transmit genetic traits: i.e. the genes.

Puzzled? So were they for a while. Ironically, the answers came not from geneticists, but from the medical community.

Pneumonia is caused by the bacterium pneumococcus. Only certain strains of pneumococci are disease causing or virulent. The difference between virulent and non-virulent strains is a hereditary property of the strain, and a major question was the chemical basis for the biological specificity of virulence.

Two medical researchers studying pneumococcus provided the discoveries identifying DNA as the hereditary material. Both were shy, meticulous, and slight of build.

One, Fred Griffith, who worked at the Ministry of Health in London, was a taxonomist who devoted his career to developing reliable techniques for classifying pathogens (any organisms that cause disease).

The second, Oswald Avery, was the son of a mystical English pastor who immigrated first to Halifax, Nova Scotia, and later to New York. Avery worked at the Rockefeller Institute near the laboratory of the famous DNA chemist P.A.T. Levene.

Griffith, working in London, found that non-virulent strains formed **rough** colonies on agar plates. Virulent strains formed **smooth** colonies. The appearance of the colonies was enough to distinguish the two types. To assay the virulence of a colony, Griffith injected the bacteria into mice.

Griffith found that heat treatment of virulent bacteria killed them. They lost their ability to cause disease. But surprisingly, after injection of killed virulent (smooth) bacteria mixed with living non-virulent (rough) pneumococci, the mice died. Furthermore, the bacteria isolated from the diseased mice were the smooth virulent type, although the only living bacteria injected were rough non-virulent. Rough bacteria were converted to smooth by a non-living extract of the smooth bacteria. Most important, the change was a permanent, inheritable one. The gene which determined rough colonies versus smooth colonies had been transferred to the recipient.

Avery, excited by this discovery of imparting a hereditary change on pneumococci, sought to identify the component of the killed smooth pneumococcus which conferred the virulent type. The difference between the smooth and rough colonies, and the virulent versus the non-virulent strains, provided assays for the change.

NON-VIRULENT (ROUGH COLONIES)

VIRULENT (SMOOTH COLONIES)

HEAT-KILLED SMOOTH

Avery's lab adapted Griffith's mixed injection procedure so it worked under more defined laboratory conditions. Living rough pneumococci were mixed with heat-killed

SERUM (ANTI-ROUGH)

smooth pneumococci in a rich bacterial growth medium. Anti-pneumococcus serum prepared against the non-virulent rough strain was added.

The living rough bacteria which he added to the broth could not survive in the presence of this serum. After several growth cycles in a test tube with the serum, pneumococci of the smooth type were detected. The transformation had worked without the need to inject the components into the mouse. The way was clear to purify the factor active in transformation, which they called "transforming principle." The highly purified transforming principle was **DNA**.

Avery's conclusions were published together with Colin MacLeod and Maclyn McCarty's in a muted paper in 1944. For years, two influential scientists at Avery's own institution, Rockefeller Institute, questioned Avery's conclusions (see page 29). P.A.T. Levene's Tetranucleotide Hypothesis was incompatible with an informational role for DNA. And Alfred Mirsky, expert on chromatin, maintained that protein contaminants of transforming principle could be the true transforming agents – not DNA!

Avery's reticence raised doubts that he recognized the importance of his own findings. For many, however, the identification of the transforming principle as DNA was an exhilirating discovery.

DNA is like a long string of beads in which each bead can be one of four kinds. Information is coded in the order in which the beads are arranged on the string. The untied part of each bead is known as a "base" and has a name: **adenine**, **guanine**, **cytosine**, or **thymine**. (In RNA, whose function we will study later, thymine is replaced by uracil.) By 1900 all of these bases were known to chemists, and were classified into two groups: the **purines**, adenine and guanine; and the **pyrimidines**, cytosine, thymine, and uracil. These are abbreviated A, G, C, T, and U.

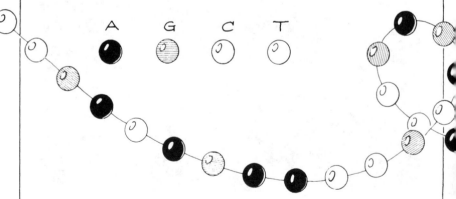

A   G   C   T

__If__ DNA carried genetic information, the ratio of the bases would probably vary. If there were exactly the same number of adenines as guanines, cytosines, and thymines, DNA might not carry information.

In the early twentieth century, chemists did not know that the genetic information resides in the **linear** arrangement of bases. Using rather crude chemical analyses, scientists were misled to believe that DNA contained exactly **equal** amounts of the four bases. This is called the Tetranucleotide Hypothesis, which originated with a German chemist, Albrecht Kossel, but it is more closely associated with the name of a Russian-born chemist who did most of his work at the Rockefeller Institute in New York, P.A.T. Levene.

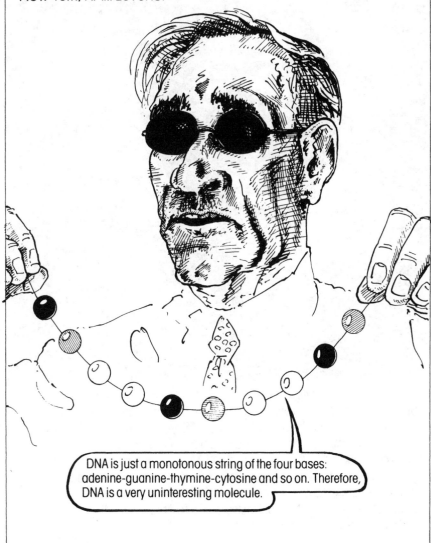

DNA is just a monotonous string of the four bases: adenine-guanine-thymine-cytosine and so on. Therefore, DNA is a very uninteresting molecule.

Levene has been much maligned for accepting the Tetranucleotide Hypothesis and even blamed for holding back genetic research for several decades as a consequence! Let's see how history has proved Levene to have been mistaken...

There are two different kinds of nucleic acids – DNA and RNA. Since, in the late nineteenth century, the primary source of what would be later called DNA was calf thymus, and what would later be identified as RNA was yeast, it was believed that DNA was only found in animals and RNA was only found in plants (such as yeast). This was an unfortunate misapprehension and it persisted in spite of the fact that RNA was frequently found in animal cells. Scientists dismissed this fact arguing that the animals had ingested plants!

At the chemical level, DNA and RNA are distinguished by: 1) that the base thymine is in DNA and is replaced by uracil in RNA; 2) that the sugar "backbones" are different. In both sugars there are five carbons arranged in a five-sided-ring structure. Such sugars are called pentoses, specifically riboses.

RNA is an abbreviation of Ribo Nucleic Acid
DNA is an abbreviation of Deoxyribo Nucleic Acid

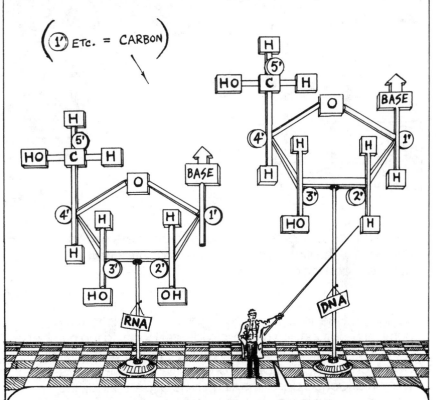

We give each sugar carbon a number; you will notice in the models that the number 2' carbon atom has a different side group in the DNA sugar as opposed to the RNA sugar. The DNA sugar lacks an oxygen atom at position number 2.
Since the RNA sugar is called "ribose" the DNA sugar is called 2'**deoxy**ribose, that is, a ribose without oxygen at atom number 2'. **That is the origin of the abbreviated forms DNA and RNA.**

By 1935 Levene showed that the sugars are connected to each other in both DNA and RNA by a phosphodiester bond, that is, through a phosphorus surrounded by oxygens:

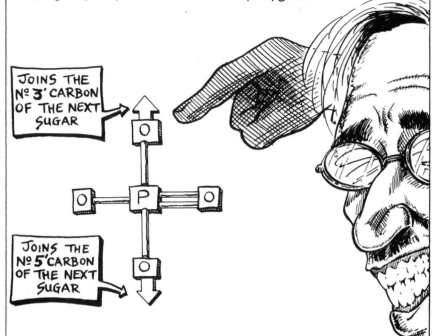

JOINS THE Nº 3' CARBON OF THE NEXT SUGAR

JOINS THE Nº 5' CARBON OF THE NEXT SUGAR

When attached together, a sugar and base are called a nucleotide. The bases (A, C, G, T or U) are attached to the number 1' carbon of the sugar.

Notice that the linkage of one sugar to the next is from the number 5' carbon, via a phosphodiester bond, to the number 3' carbon of the next sugar. This is why DNA is like a long set of beads on a string. The beads on the string are the sugars, each with an attached base. The string is held together by the phosphodiester bonds between sugars.

Because the phosphate linkage between sugars runs from the number 5' carbon to the number 3' (as shown on the next page), we can say that the sugar backbone can be oriented in space. Biochemists talk of "moving in the 5' to 3' direction" down a DNA chain.

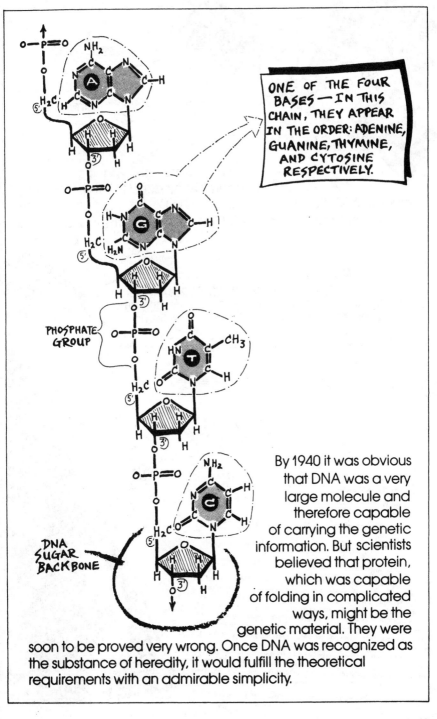

ONE OF THE FOUR BASES — IN THIS CHAIN, THEY APPEAR IN THE ORDER: ADENINE, GUANINE, THYMINE, AND CYTOSINE RESPECTIVELY.

PHOSPHATE GROUP

DNA SUGAR BACKBONE

By 1940 it was obvious that DNA was a very large molecule and therefore capable of carrying the genetic information. But scientists believed that protein, which was capable of folding in complicated ways, might be the genetic material. They were soon to be proved very wrong. Once DNA was recognized as the substance of heredity, it would fulfill the theoretical requirements with an admirable simplicity.

THE NEW BIOLOGY

The immediate question was how DNA could be informational.

Erwin Chargaff, a young Viennese-trained chemist, unlocked the first chemical clues to genetic information storage in DNA. Chargaff read the Avery, MacLeod, and McCarty paper on the transforming principle.

I see in dark contours the beginning of a grammar of biology ... Avery gave us the first text of a new language, or rather he showed us where to look for it. I resolved to search for this text.

'DIES ON THE CHEMICAL NATURE OF THE SUBSTANCE
NDUCING TRANSFORMATION OF PNEUMOCOCCAL TYPES

UCTION OF TRANSFORMATION BY A DESOXYRIBONUCLEIC ACID FRACTION
ISOLATED FROM PNEUMOCOCCUS TYPE III

BY OSWALD T. AVERY, M.D., COLIN M. MACLEOD, M.D., AND
MACLYN McCARTY,* M.D.

(From the Hospital of The Rockefeller Institute for Medical Research)

PLATE 1

(Received for publication, November 1, 1943)

Biologists have long attempted by chemical means to induce in higher
ganisms predictable and specific changes which thereafter could be trans-
tted in series as hereditary characters. Among microörganisms the most
iking example of inheritable and specific alterations in cell structure and
nction that can be experimentally induced and are reproducible under well
fined and adequately controlled conditions is the transformation of specific
pes of Pneumococcus. This phenomenon was first described by Griffith (1)
ho succeeded in transforming an attenuated and non-encapsulated (R)
riant derived from one specific type into fully encapsulated and virulent (S)
lls of a heterologous specific type. A typical instance will suffice to illustrate
e techniques originally used and serve to indicate the wide variety of trans-
rmations that are possible within the limits of this bacterial species.

. . . . . . mice injected subcutaneously with a small amount of a living

34

Chargaff's approach was to use the methods of quantitative analysis, bolstered by newly available techniques for separating the four bases. He purified DNA samples and then carefully quantified the amount of the four bases, A, G, C, and T.

The solvent systems and the visualization of the separated spots were primitive, but we could separate and identify as little as five micrograms of each substance.

When Chargaff measured the base compositions of DNA from many sources he noted regularities summarized in "Chargaff's Rules":

(a) THE QUANTITY OF A + G =
    "   "   " C + T.
(b) A CONTENT ALWAYS = T CONTENT
    G   "   "   = C   "

This chemical symmetry was at once intriguing and enigmatic. It suggested that there was an underlying regularity to the composition of DNA. But Chargaff's Rules on their own were insufficient to explain the regularities that Chargaff observed.

DNA is a macromolecule and most of its interesting features are lost when it is degraded. The successful experimental approach for determining DNA structure had to be capable of analyzing DNA intact, in its macromolecular form. One such technique was X-ray diffraction, which was developed in Cambridge by the Braggs, father and son, at the Cavendish Laboratory.

SIR LAWRENCE BRAGG
CAVENDISH PROFESSOR
1890 – 1971

In X-ray diffraction, a fine beam of X-rays is passed through a crystal of the substance whose structure is under analysis. It interacts with the atoms in the crystal, and re-emerges as a complex pattern of beams that may be captured on X-ray film. By analyzing changes in the beam imparted by the specimen, the structure of the unknown molecule may be deduced.

The Cambridge lab sought a daring application of X-ray diffraction to the very complex biological macromolecules, the proteins. Max Perutz, an Austrian, was enlisted to lead the Cambridge protein structure team.

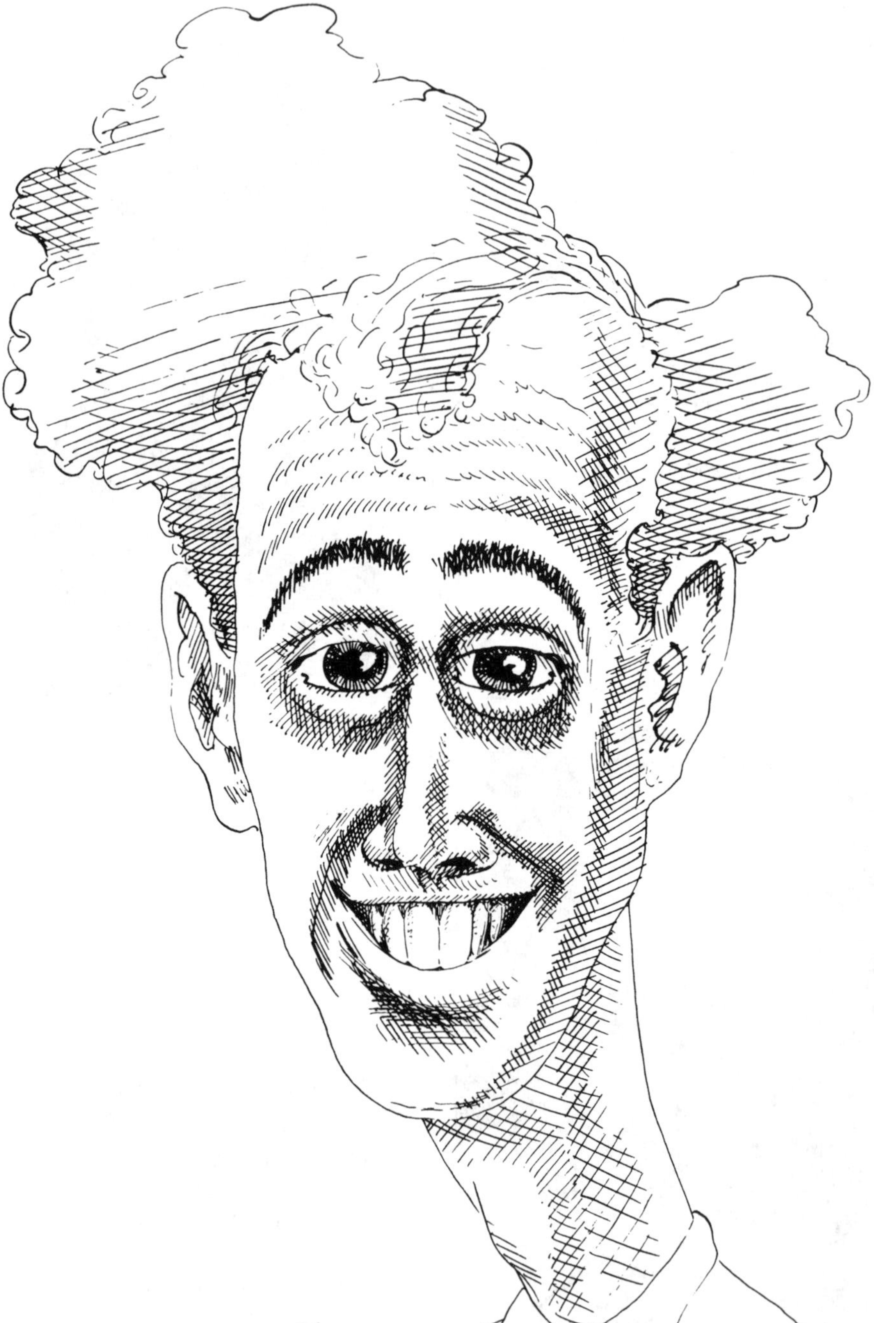
James D. Watson

Perutz's team did not consist solely of investigators interested in proteins. The Cambridge lab was joined in 1951 by an American, James D. Watson, who had other questions on his mind. It was Watson's interest in genes which led him to the Cavendish. As an undergraduate at the University of Chicago, Watson divided his time between bird watching and musing about biology. He had the good fortune of studying with Salvador Luria, a founder of the phage group, from whom he learned the principles of phage genetics.

Watson appreciated that DNA was the molecular key to genetics, but his weak background in chemistry limited his ability to understand genetic phenomena in terms of the chemical properties of the DNA molecule. On completing studies with Luria, Watson journeyed to Copenhagen, where he pursued phage studies. Watson enjoyed journeying from his Copenhagen laboratory base, and in 1951 he travelled to a conference in Naples, where he chanced to meet Maurice Wilkins, a London crystallographer interested in DNA.

At the Naples meeting, Wilkins briefly showed an X-ray photo of DNA. Unlike a traditional photograph obtained with a camera lens and daylight, simply looking at the X-ray photo did not disclose the structure of the DNA molecule. Even the sharp spots easily obtained with simple mineral specimens were missing.

However, the presence of a regular, if fuzzy, geometric pattern in the X-ray photo confirms that the DNA sample is, at least partially, crystalline.

I would conclude that genes must have some regularity to their structure which would allow them to pack together in a nearly crystalline arrangement. The regularity could simplify the deduction of the structure of a gene.

Inspired by the X-ray photo, Watson sought a lab where he too could delve into the chemical structure of DNA. Watson arranged a shift from Copenhagen to Cambridge where he joined Max Perutz's group. Protein crystallography did not come easily to Watson, and he soon found himself preoccupied by conversations with Francis Crick.

Francis Crick was a 35-year-old physicist who was developing the mathematics of X-ray diffraction for application to macromolecules. At heart, Crick was a theorist, trained in physics but drawn to biology by a fascination for understanding the activities of living things through . . .

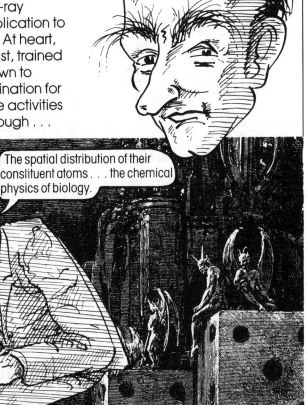

The spatial distribution of their constituent atoms . . . the chemical physics of biology.

Crick studied physics until the outbreak of World War II, and then served in the Admiralty designing ingenious magnetic mines. In 1949, he moved to the Cavendish group. Crick soon taught himself X-ray diffraction theory, and the current state of the protein structure problem.

From their first encounter in 1951, Watson and Crick thrived on each other's discussions. They agreed that the solution of DNA structure was the paramount problem of genetics. But Crick the theorist, and Watson the untutored newcomer, could contribute little new information of their own.

Outside Cambridge, one other scientist had novel insight into protein structure. Linus Pauling of Cal Tech, chemistry genius, proposed that protein chains fold in helical form. Crick knew this model well, and from studying Pauling's proposals he learned the theory of diffraction of helical macromolecules.

The best X-ray data, in part presented at the Naples meeting by Maurice Wilkins, resided in London. Wilkins was also a physicist who turned to biophysics, and in 1950, together with his graduate student Raymond Gosling, Wilkins obtained good X-ray patterns from DNA fibers.

Generally, the most definitive X-ray patterns are obtained when the specimen is crystalline. DNA, a long thread-like molecule, could be pulled into fibers, in which individual DNA molecules oriented themselves side by side, stretched out parallel to one another. Although not truly crystalline, the DNA fibers had sufficient order that informative X-ray patterns could be obtained. In forming the first fibers, most of the water was removed from the DNA, and the resulting structure was called "the A form" of DNA.

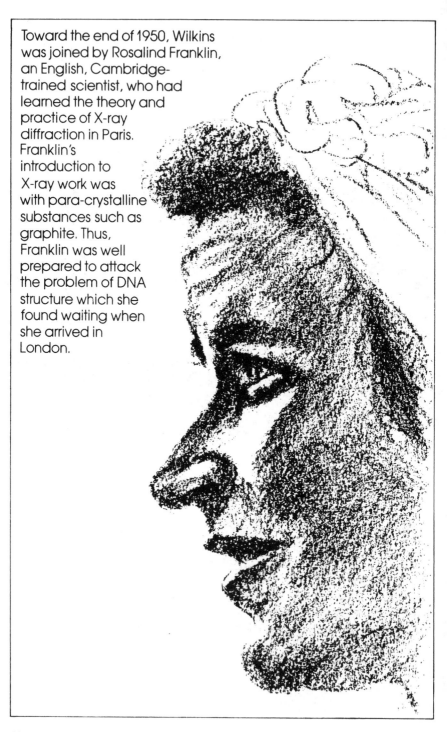

Toward the end of 1950, Wilkins
was joined by Rosalind Franklin,
an English, Cambridge-
trained scientist, who had
learned the theory and
practice of X-ray
diffraction in Paris.
Franklin's
introduction to
X-ray work was
with para-crystalline
substances such as
graphite. Thus,
Franklin was well
prepared to attack
the problem of DNA
structure which she
found waiting when
she arrived in
London.

Franklin was dedicated to her work, fully competent, eager to solve the DNA problem, and not disposed to be deferential to Wilkins. During the time they shared a laboratory, Franklin and Wilkins remained distant, with Wilkins reserved and somewhat formal. Franklin, although isolated, was professionally assertive in her studies of DNA.

45

By the autumn of 1951, Franklin had her first success. She devised an improved method for adding water back to the A form of DNA fibers. When hydrated, the DNA underwent a dramatic structural change, observable by the diffraction technique. In November 1951, Franklin gave the first public presentation of her results to a small gathering at Kings College. In the audience of this seminar was Watson.

"The results suggest a helical structure (which must be very closely packed) containing probably 2, 3 or 4 co-axial nucleic acid chains per helical unit, and having the phosphate groups near the outside."

"Naturally I was delighted when Maurice said I would be welcome at Rosy's talk. For the first time I had a real incentive to learn some crystallography. I did not want Rosy to speak over my head."

After the meeting, Watson and Wilkins had a Chinese dinner together. Watson left with the impression that Franklin had only refined Wilkins's existing data, and might actually slow the investigation because of her distant relationship with Wilkins. Watson returned to Cambridge and related his recollection of Franklin's talk to Crick.

Crick was tantalized by the possibility that the data already available might limit the structures for DNA to a small number of possibilities. The structure might be deduced by proposing a hypothetical structure, and fitting the experimental data against the predictions of the model.

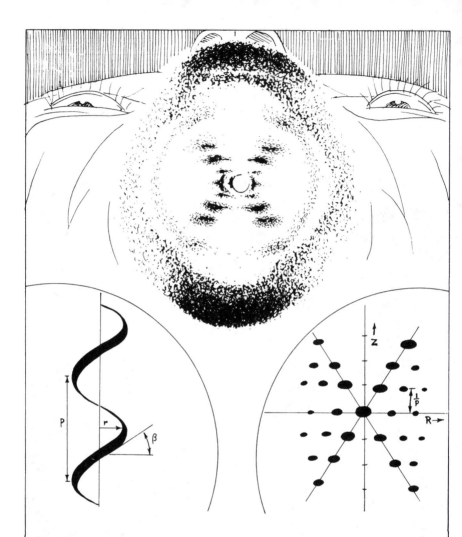

The X-ray diffraction pattern was in the form of a "Maltese cross" characteristic of a helical molecule. Franklin and Wilkins had already recognized the probable helicity of DNA. Fortunately for Crick, the alpha helix protein structure proposed by Pauling had spurred an intense review and re-derivation of the theory of helical molecule diffraction, which Crick made together with the Scottish statistician William Cochran. There were many possible helices: two stranded, three stranded, four stranded – and each could wind with a range of pitches and diameters. Many of the fundamental dimensions could be deduced from the X-ray photos.

In 1952 Erwin Chargaff traveled to Cambridge, and at the insistence of Perutz's co-worker, John Kendrew, he spoke to two people, Watson and Crick, at the Cavendish Laboratory who were "trying to do something with nucleic acids." Chargaff wrote of that meeting ...

it there. I have to point out that mytholog-ical or historical couples—Castor and Pollux, Harmodios and Aristogeiton, Romeo and Juliet—must have appeared quite differently before the deed than after. In any event I seem to have missed the shiver of recognition of a historical moment: a change in the rhythm of the heart-beats of biology. So far as I could make out they wanted, unencumbered by any knowledge of the chemistry involved, to fit DNA into a helix.

"The high point of Chargaff's scorn," wrote James D. Watson of this meeting, "came when he led Francis into admitting that he did not remember the chemical difference among the four bases." Despite the scorn, Chargaff related in detail his findings about DNA base ratios. He described the chemical symmetry of Chargaff's Rules.

I am undeterred by Chargaff's scorn. But each visitor brings word of new facts about DNA. We must solve its structure!

Great! Let's build another model...

DID CHARGAFF SAY A ATTRACTS G OR G ATTRACTS T?

A difficult test of any model would be an explanation of the novel ability of genes to duplicate themselves. Watson was aware of a hypothesis that gene duplication relied upon formation of "complementary surfaces," from which a new gene could be constructed.

The mechanism would be similar to preparing a mold of an object, from which a replica of the original object could be cast.

An alternative scheme for duplication was direct copying, with no complementary intermediate.

Meanwhile, Pauling had turned his attention from proteins and proposed an unworkable structure for DNA. Watson recognized that Pauling's model failed to account for the acidic nature of DNA. Stabilizing forces, critical to Pauling's model, probably didn't exist. Watson was convinced that Pauling would soon be aware of his error, and then would intensify his effort to derive the correct structure.

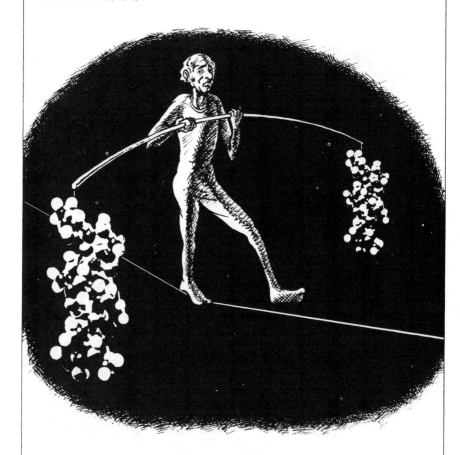

During this time, Franklin pressed forward the X-ray studies of the B form of DNA. Watson was privy to her progress by virtue of periodic meetings with Maurice Wilkins in London. When the Cochran-Crick diffraction theory was used to test the B form patterns, it was evident that hydrated DNA was also a helix.

To bring together disparate evidence for the structure, Watson wanted to build precise representations of DNA helices. Machinists at the Cavendish were asked to make metal replicas of the purine and pyrimidine bases. In some models, the phosphate backbone was on the interior, as with Pauling's model. In others, it was on the exterior. Watson tried groupings of bases which could form hydrogen bonds and stabilize the helix. At one point, Watson considered a model in which A paired with A, T with T, and so on, such that "like paired with like."

However, the like-like model was soon rejected because Watson had employed the wrong chemical forms of T and G.

Watson continued to shuffle the cut-outs of the bases in different combinations.

Suddenly I became aware that an adenine-thymine pair was identical in shape to a guanine-cytosine pair held together by at least two hydrogen bonds. All the hydrogen bonds seemed to form naturally, no fudging was required to make the two types of base pairs identical in shape.

Most important was that the two types of base pairs had the same overall size and shape. Thus, in fitting these pairs into the helix, any order of A-T's and G-C's could be accommodated, and the same regular exterior phosphate backbone could be maintained.

Together, Watson and Crick assembled a three-dimensional structure of a double helix which contained the newly conceived base pairs and employed the dimensions derived from X-ray measurements.

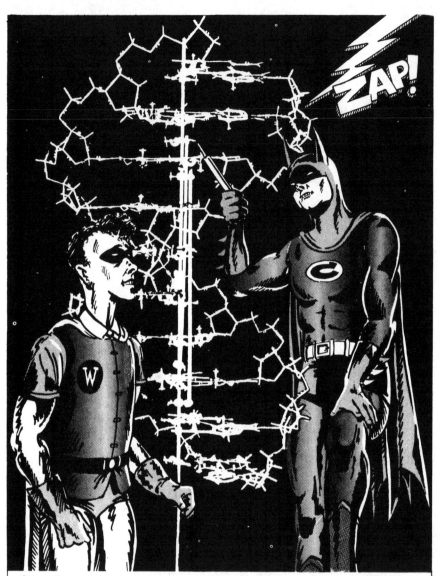

Standing as high as a man, the model had brass bases and wire sugars, and was held together by screws. Strikingly, the model at once suggested a mechanism for replicating genes. The base sequence of one chain automatically determined the sequence of the other. Also, the A-T and G-C base pairs immediately explained Chargaff's Rules. With a C facing every G, and a T opposite every A, the A-T and G-C equivalences were ensured.

The model was submitted to *Nature* in a 900-word manuscript, together with two other separate reports, one from Wilkins and one from Franklin. Watson's sister typed the final draft of the historic manuscript, which was submitted in April 1953, when Watson was 25 years old.

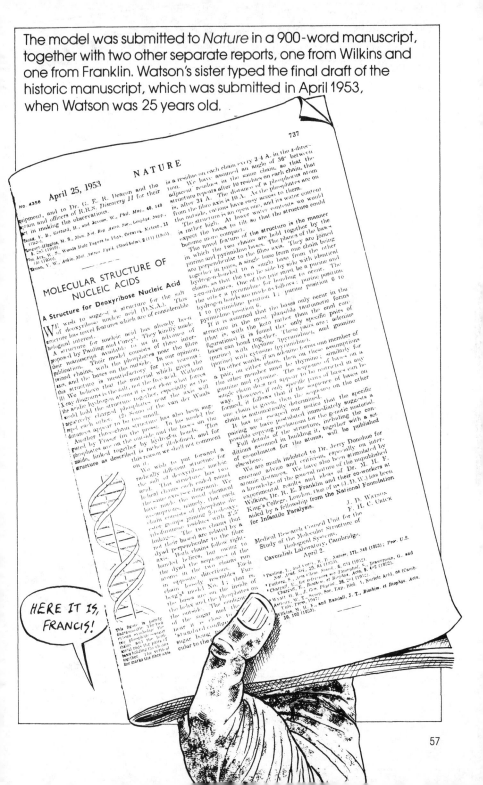

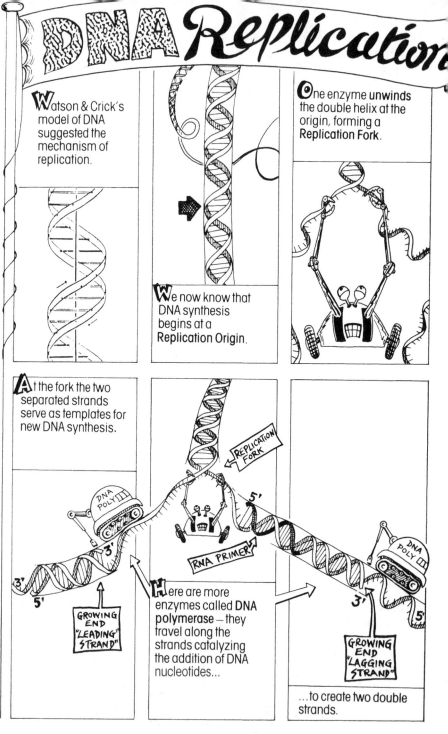

DNA Replication

Watson & Crick's model of DNA suggested the mechanism of replication.

We now know that DNA synthesis begins at a **Replication Origin**.

One enzyme **unwinds** the double helix at the origin, forming a **Replication Fork**.

At the fork the two separated strands serve as templates for new DNA synthesis.

REPLICATION FORK

RNA PRIMER

DNA POLY III

5'

3'

3'

5'

GROWING END "LEADING STRAND"

Here are more enzymes called **DNA polymerase** — they travel along the strands catalyzing the addition of DNA nucleotides...

DNA POLY III

3'

5'

GROWING END "LAGGING STRAND"

...to create two double strands.

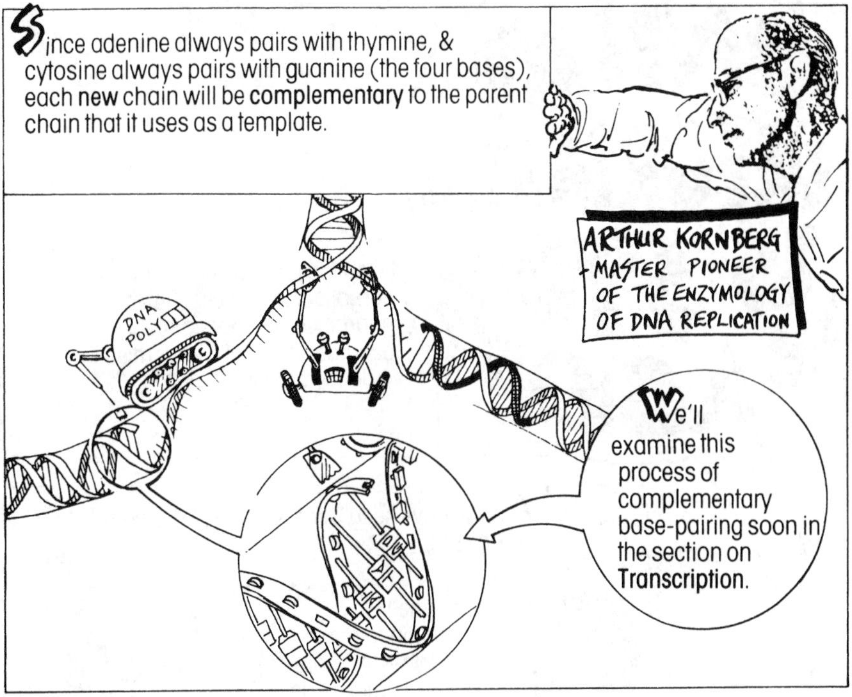
Since adenine always pairs with thymine, & cytosine always pairs with guanine (the four bases), each **new** chain will be **complementary** to the parent chain that it uses as a template.
ARTHUR KORNBERG
- MASTER PIONEER OF THE ENZYMOLOGY OF DNA REPLICATION
DNA POLY III
We'll examine this process of complementary base-pairing soon in the section on **Transcription**.

The interaction of these and many other features near the replication fork results in two new double helices. Each one has **one** chain from the original DNA molecule & one chain that has been newly formed.

Here's a representation of what DNA replication looks like when hugely magnified.

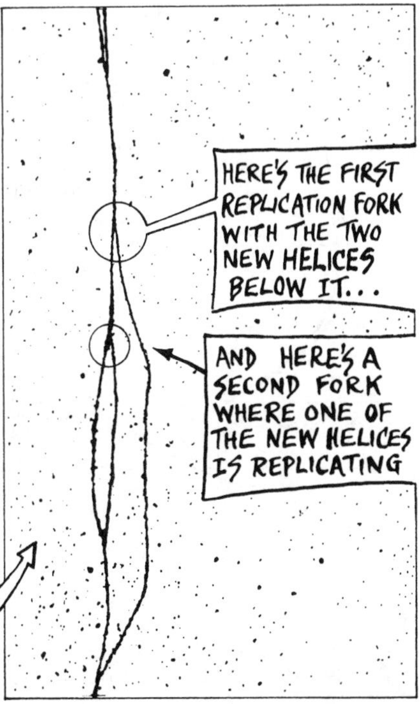
HERE'S THE FIRST REPLICATION FORK WITH THE TWO NEW HELICES BELOW IT...
AND HERE'S A SECOND FORK WHERE ONE OF THE NEW HELICES IS REPLICATING

WHAT INFORMATION IS STORED IN A GENE?

Replication is an extremely complicated process, but this guarantees the near perfect accuracy of genetic transmission & consequently, life itself!

As we have seen **George Beadle** & **Edward Tatum** proposed:

(SEE PAGE 23)

Each gene has information to make one enzyme, or more precisely to make one protein.

The critical clue to the complexity of this genetic information came from **Fred Sanger**, an English biochemist, who determined the complete amino acid sequence of the hormone insulin.

Insulin is a protein. Proteins are long chains of amino acids. There are twenty amino acid types.

Sanger proved that proteins have specific structures. This had implications for genes.

1 Sanger sequenced insulin by specifically degrading it into short fragments which were separated by a procedure known as "**fingerprinting**."

2

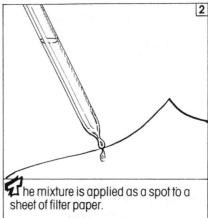

The mixture is applied as a spot to a sheet of filter paper.

3 Solvent is passed in one direction & electric current in the perpendicular direction.

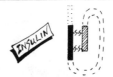

4

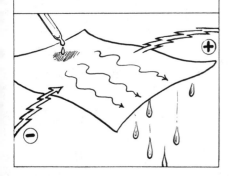

Depending on their solubility & electric charge, different fragments are moved to different positions on the paper, creating a distinct pattern.

5 When Sanger inadvertently touched the paper sheets before visualizing the protein fragments, spots appeared that were caused by protein from his fingertips, according to one story.

He called the patterns "fingerprints" because each protein produced a unique pattern of spots.

6

Like fingerprints, the patterns were characteristic for each protein: simple & reproducible. Sanger concluded that insulin had a specific structure. He next reassembled the short sequences into longer ones, & deduced the complete structure of insulin.

The striking conclusion for genetics deduced by Sanger was that the protein insulin had a precisely defined amino acid sequence.

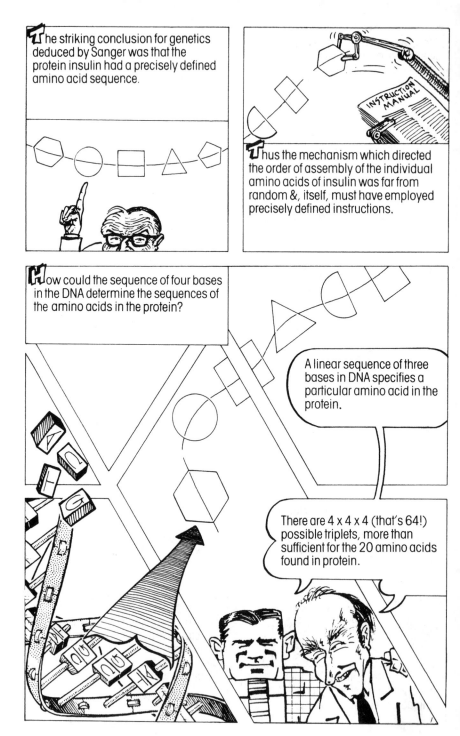

Thus the mechanism which directed the order of assembly of the individual amino acids of insulin was far from random &, itself, must have employed precisely defined instructions.

How could the sequence of four bases in the DNA determine the sequences of the amino acids in the protein?

A linear sequence of three bases in DNA specifies a particular amino acid in the protein.

There are 4 x 4 x 4 (that's 64!) possible triplets, more than sufficient for the 20 amino acids found in protein.

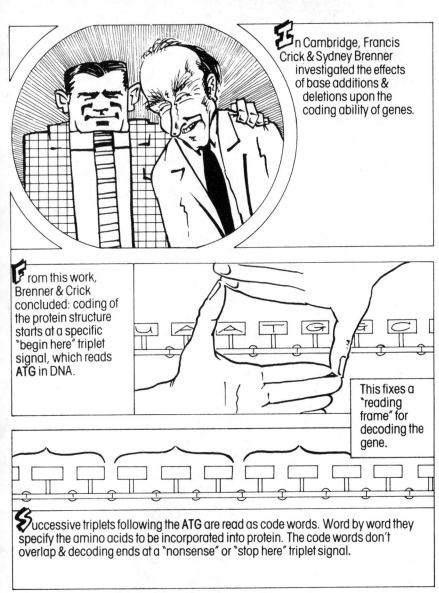

In Cambridge, Francis Crick & Sydney Brenner investigated the effects of base additions & deletions upon the coding ability of genes.

From this work, Brenner & Crick concluded: coding of the protein structure starts at a specific "begin here" triplet signal, which reads **ATG** in DNA.

This fixes a "reading frame" for decoding the gene.

Successive triplets following the **ATG** are read as code words. Word by word they specify the amino acids to be incorporated into protein. The code words don't overlap & decoding ends at a "nonsense" or "stop here" triplet signal.

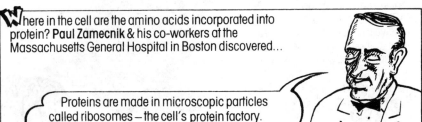

Where in the cell are the amino acids incorporated into protein? **Paul Zamecnik** & his co-workers at the Massachusetts General Hospital in Boston discovered...

Proteins are made in microscopic particles called ribosomes — the cell's protein factory.

But what is happening inside the ribosome

F.H.C. CRICK

Francis Crick had a brainstorm!! At the time he and a group of twenty other scientists had formed an elite **RNA Tie Club**. It was to this exclusive membership — and to them only — that Crick sent a mimeographed copy of his brainstorm:

THE ADAPTOR HYPOTHESIS

Crick's Adaptor Hypothesis dealt with the problem of how the code held in the DNA double helix gets **translated** into protein.

I KNOW (OR CAN ASSUME) THE FOLLOWING

The code in DNA is **linear**. Its sequence of bases is like a string of beads.

Every group of three bases in DNA codes for **one** amino acid.

A protein is made up from a string of amino acids (of which there are twenty different kinds).

The amino acids in a protein are in the same sequence as their codes in DNA.

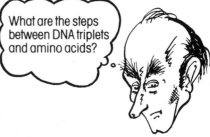

What are the steps between DNA triplets and amino acids?

PART WHICH RECOGNIZES A TRIPLET OF BASES

PART WHICH CARRIES ONE OF THE TWENTY AMINO ACIDS

There must be an **Adaptor** molecule which has one side to recognize one amino acid & the other to recognize its triplet code.

I t wasn't long before scientists discovered the adaptors Crick had predicted. They were made out of RNA & **translated** the genetic code (which was stored in a molecule called **messenger** RNA) inside the **ribosome**. The ribosome was the assembly line for the production of proteins.

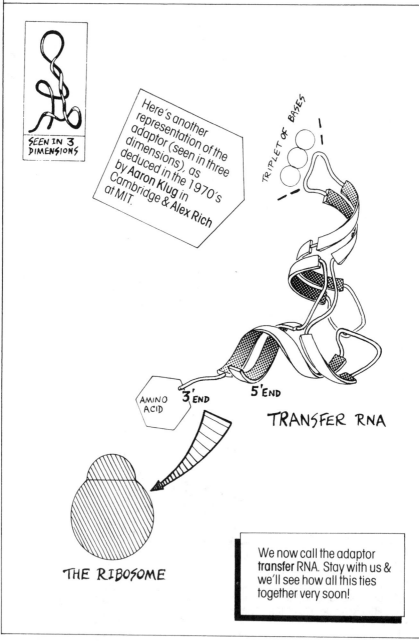

SEEN IN 3 DIMENSIONS

Here's another representation of the adaptor (seen in three dimensions), as deduced in the 1970's by **Aaron Klug** in Cambridge & **Alex Rich** at MIT.

TRIPLET OF BASES

AMINO ACID

3' END

5' END

TRANSFER RNA

THE RIBOSOME

We now call the adaptor **transfer** RNA. Stay with us & we'll see how all this ties together very soon!

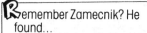

Remember Zamecnik? He found…

Proteins are made in the ribosomes & there is **no** DNA in the ribosomes.

How did the coded information get from the cell's DNA to the ribosome?

It was only after **Arthur Pardee, François Jacob & Jacques Monod** working together at the Pasteur Institute in Paris performed their famous "**PaJaMo**" experiments (named after themselves), that this piece of the puzzle fell into place.

β-G.

(SEE PAGE 88)

The gene for making an enzyme, beta-galactosidase (which digested the sugar lactose), were transferred from the male bacteria to the females which were not capable of making the enzyme.
(As we shall see later, bacteria have "sex.")

(SEE PAGE 95)

The gene for beta-galactosidase no sooner entered the female than it (she) began producing the enzyme beta-galactosidase.

This surprised most scientists because it was widely assumed that before a gene could be expressed (before a cell could produce the protein for which the gene codes) **stable** cellular structures would have to form & accumulate.

This meant that there would be a delay before the gene products appeared in a bacterium.

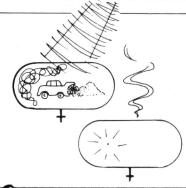

Other experiments were performed in which a newly inserted gene (DNA) was destroyed by a radioactive technique (which we needn't detail). Once the gene was destroyed the bacterium stopped producing the particular product encoded in that gene. This too was surprising.

After much discussion on Good Friday in 1960 in Cambridge, England, Crick, Brenner & Jacob concluded that these experiments showed that the template for protein synthesis (what the tRNA attached to) was **not stable**.

RHUBARB RHUBARB

They concluded that as soon as the male DNA entered the female bacterium **unstable, short-lived** copies of the DNA were formed, & these copies were the templates for protein synthesis.

The center for protein assembly — the ribosome — was known to be stable...

and tRNA was also known to be stable...

— so a **Messenger RNA** was postulated.

DNA SUGAR

RNA SUGAR

mRNA

RNA because:
1. RNA & **not** DNA was found in the ribosomal protein factories;
2. An RNA copy of the DNA, generated by base-pairing, would store the same information as the DNA itself.

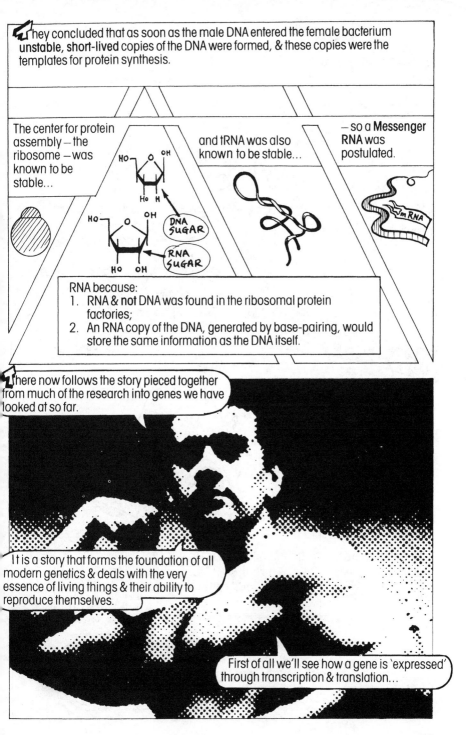

There now follows the story pieced together from much of the research into genes we have looked at so far.

It is a story that forms the foundation of all modern genetics & deals with the very essence of living things & their ability to reproduce themselves.

First of all we'll see how a gene is 'expressed' through transcription & translation...

69

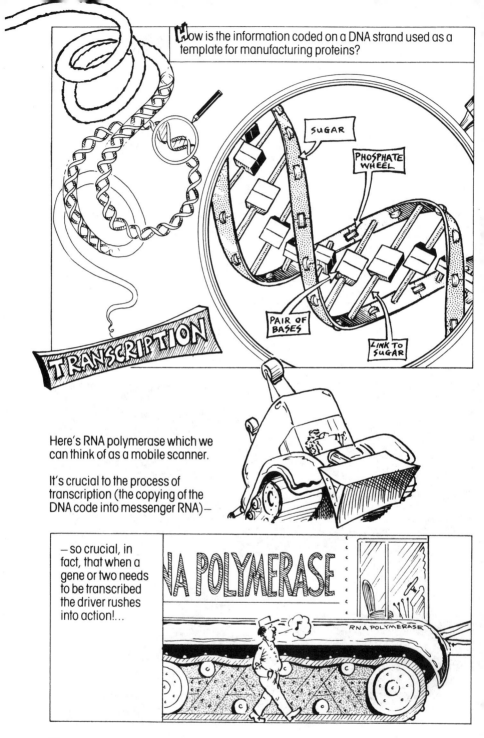

How is the information coded on a DNA strand used as a template for manufacturing proteins?

SUGAR

PHOSPHATE WHEEL

PAIR OF BASES

LINK TO SUGAR

TRANSCRIPTION

Here's RNA polymerase which we can think of as a mobile scanner.

It's crucial to the process of transcription (the copying of the DNA code into messenger RNA) —

— so crucial, in fact, that when a gene or two needs to be transcribed the driver rushes into action!...

NA POLYMERASE

RNA POLYMERASE

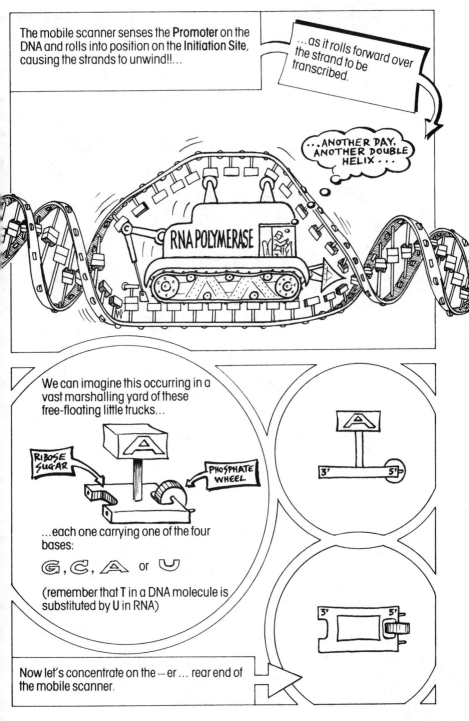

The mobile scanner senses the **Promoter** on the DNA and rolls into position on the **Initiation Site**, causing the strands to unwind!!...

...as it rolls forward over the strand to be transcribed.

...ANOTHER DAY, ANOTHER DOUBLE HELIX...

RNA POLYMERASE

We can imagine this occurring in a vast marshalling yard of these free-floating little trucks...

RIBOSE SUGAR

PHOSPHATE WHEEL

...each one carrying one of the four bases:

G, C, A or U

(remember that T in a DNA molecule is substituted by U in RNA)

Now let's concentrate on the -- er ... rear end of the mobile scanner.

71

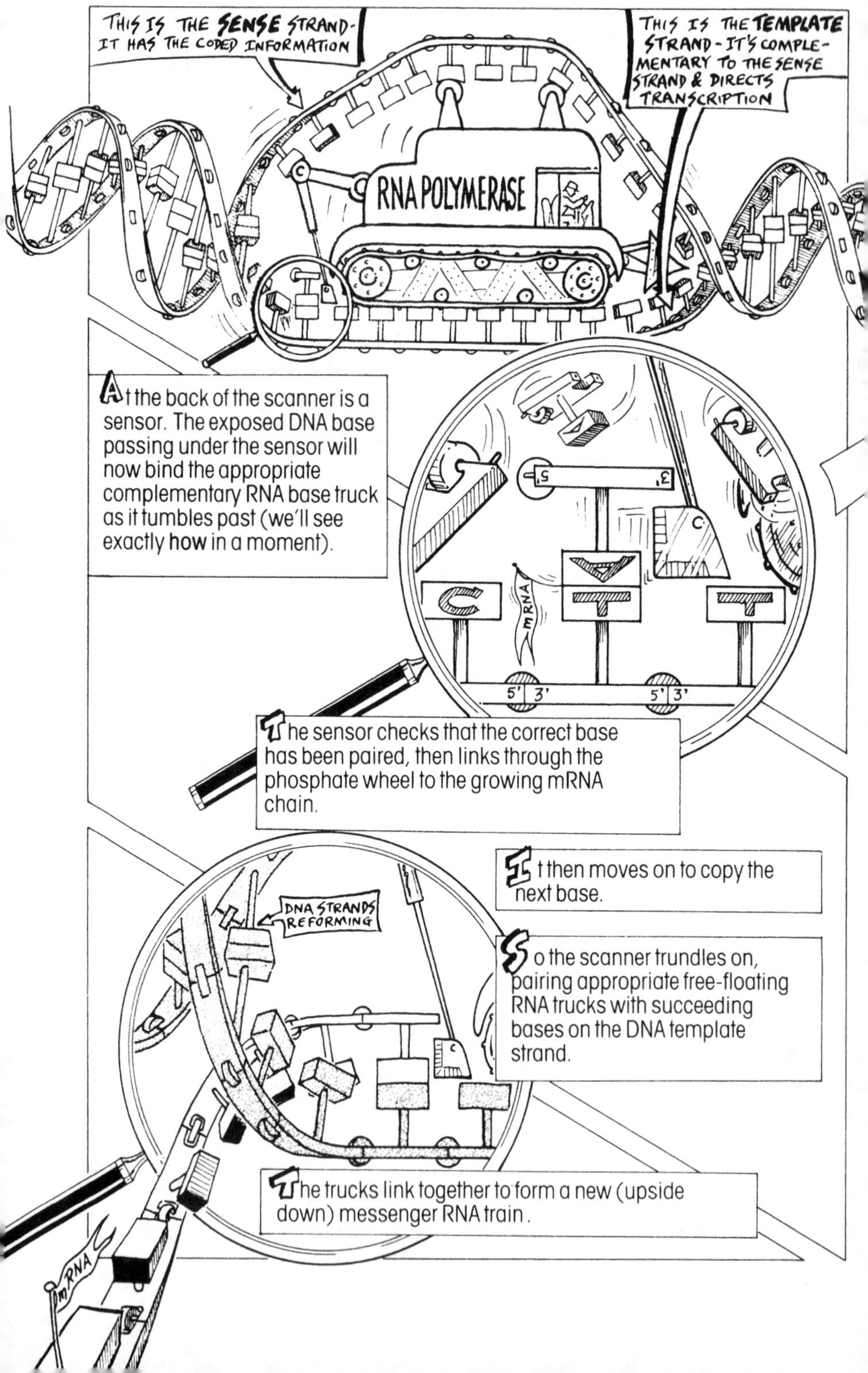
THIS IS THE SENSE STRAND- IT HAS THE CODED INFORMATION
THIS IS THE TEMPLATE STRAND - IT'S COMPLEMENTARY TO THE SENSE STRAND & DIRECTS TRANSCRIPTION
RNA POLYMERASE
At the back of the scanner is a sensor. The exposed DNA base passing under the sensor will now bind the appropriate complementary RNA base truck as it tumbles past (we'll see exactly **how** in a moment).
5'
3'
C
mRNA
T
T
5'
3'
5'
3'
The sensor checks that the correct base has been paired, then links through the phosphate wheel to the growing mRNA chain.
It then moves on to copy the next base.
DNA STRANDS REFORMING
So the scanner trundles on, pairing appropriate free-floating RNA trucks with succeeding bases on the DNA template strand.
The trucks link together to form a new (upside down) messenger RNA train.
mRNA

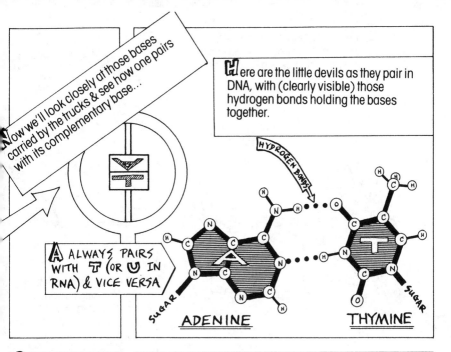

Now we'll look closely at those bases carried by the trucks & see how one pairs with its complementary base...

Here are the little devils as they pair in DNA, with (clearly visible) those hydrogen bonds holding the bases together.

HYDROGEN BONDS

A ALWAYS PAIRS WITH T (OR U IN RNA) & VICE VERSA

SUGAR

ADENINE   THYMINE

SUGAR

During transcription the incoming RNA trucks pair precisely with the DNA template strand. Because this copy is complementary to the template, it has the same sequence of bases as the DNA sense strand and therefore contains the coded genetic information.

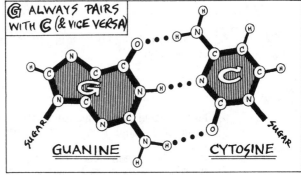

G ALWAYS PAIRS WITH C (& VICE VERSA)

SUGAR

GUANINE   CYTOSINE

SUGAR

Here's the structure of DNA
r = ribose sugar
p = phosphate

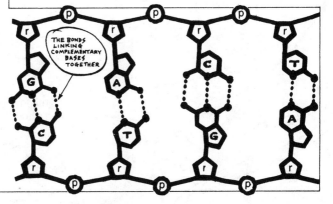

THE BONDS LINKING COMPLEMENTARY BASES TOGETHER

The mobile scanner finally completes its task, reaches the **Transcription Termination Site** on the DNA double helix, and the last truck is added to the train.

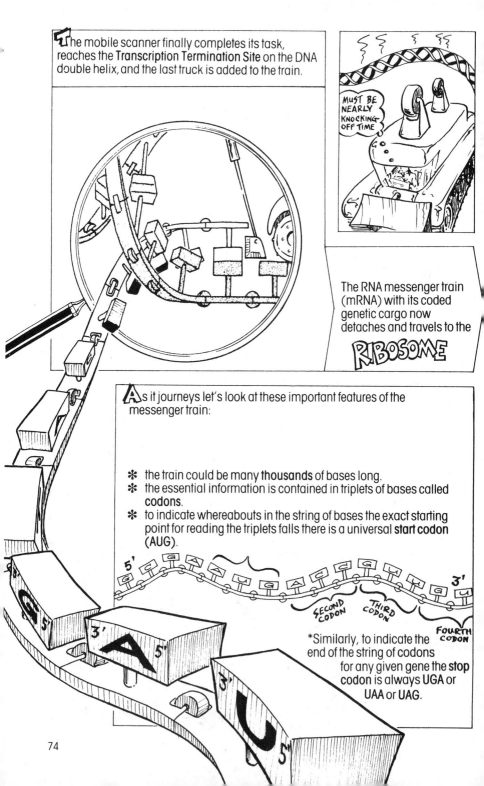

MUST BE NEARLY KNOCKING OFF TIME

The RNA messenger train (mRNA) with its coded genetic cargo now detaches and travels to the

RIBOSOME

As it journeys let's look at these important features of the messenger train:

* the train could be many **thousands** of bases long.
* the essential information is contained in triplets of bases called **codons**.
* to indicate whereabouts in the string of bases the exact starting point for reading the triplets falls there is a universal **start codon** (AUG).

SECOND CODON   THIRD CODON   FOURTH CODON

*Similarly, to indicate the end of the string of codons for any given gene the **stop** codon is always **UGA** or **UAA** or **UAG**.

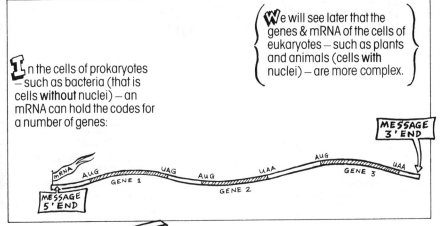

In the cells of prokaryotes – such as bacteria (that is cells **without** nuclei) – an mRNA can hold the codes for a number of genes:

We will see later that the genes & mRNA of the cells of eukaryotes – such as plants and animals (cells **with** nuclei) – are more complex.

MESSAGE 3' END

MESSAGE 5' END

mRNA

AUG

UAG GENE 1

AUG GENE 2

UAA

AUG GENE 3

UAA

TRANSLATION

RIBOSOME INC.

PROTEIN TRANSLATING FACTORY

Let's take a closer look…

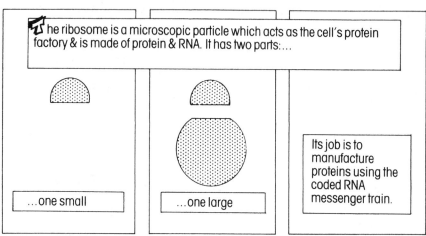

The ribosome is a microscopic particle which acts as the cell's protein factory & is made of protein & RNA. It has two parts:…

…one small

…one large

Its job is to manufacture proteins using the coded RNA messenger train.

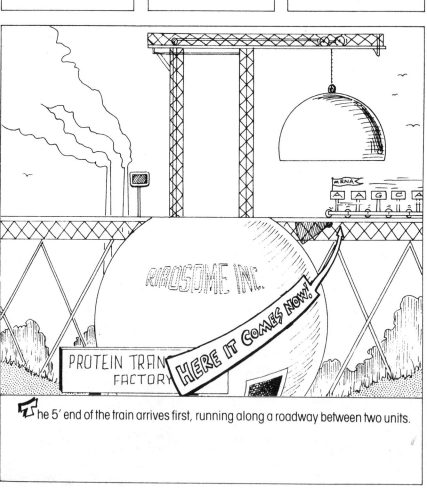

The 5' end of the train arrives first, running along a roadway between two units.

These two codons then move into position with the small unit above them, to occupy platforms P & A respectively in the large unit.

(In the bowels of the factory)

When the <u>start codon</u> passes under the small unit, it causes a gantry to lower that unit onto **AUG** & the **next codon**. At the same time an automatic ramp flips the train **upside down**!!

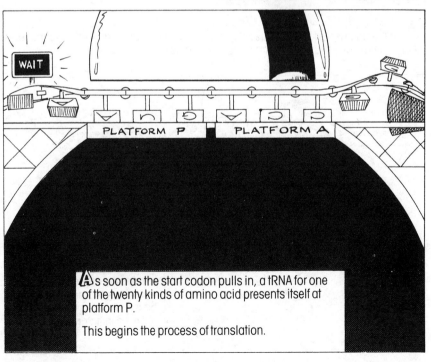

WAIT

PLATFORM P   PLATFORM A

As soon as the start codon pulls in, a tRNA for one of the twenty kinds of amino acid presents itself at platform P.

This begins the process of translation.

JUST A MINUTE! tRNA?? WHAT'S THAT??!

Remember Crick's Adaptor Hypothesis? It pointed toward a small RNA molecule (some 70 to 90 bases long) in a clover leaf shape when seen in two dimensions. It is called transfer RNA (tRNA) as its job is to transfer amino acids from a free state to a growing protein chain. After all, that's why we're all here, isn't it?

It has two main parts: a head consisting of a triplet of bases (which is called the anti-codon as it complements the codon of the mRNA train); and a trailer at the rear which carries one of the twenty amino acids (the **components** which make up **proteins**).

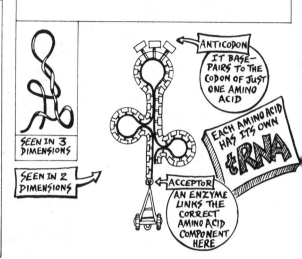

SEEN IN 3 DIMENSIONS

SEEN IN 2 DIMENSIONS

ANTICODON
IT BASE-PAIRS TO THE CODON OF JUST ONE AMINO ACID

EACH AMINO ACID HAS ITS OWN tRNA

ACCEPTOR
AN ENZYME LINKS THE CORRECT AMINO ACID COMPONENT HERE

We can imagine that beneath the protein factory is a vast underground warehouse containing many many tRNAs carrying amino acids.

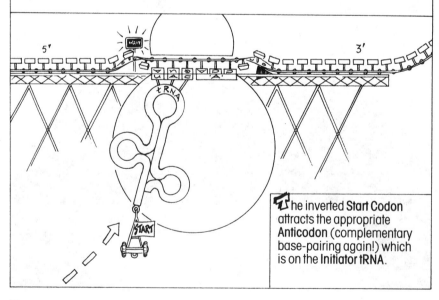

5′   WAIT   3′

tRNA

START

The inverted **Start Codon** attracts the appropriate **Anticodon** (complementary base-pairing again!) which is on the **Initiator tRNA**.

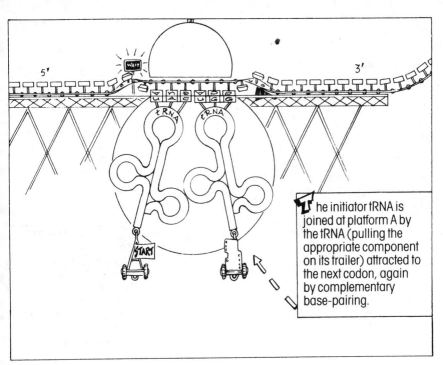

The initiator tRNA is joined at platform A by the tRNA (pulling the appropriate component on its trailer) attracted to the next codon, again by complementary base-pairing.

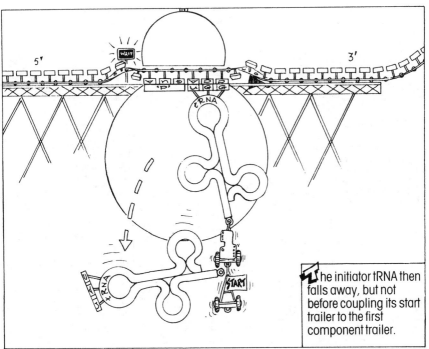

The initiator tRNA then falls away, but not before coupling its start trailer to the first component trailer.

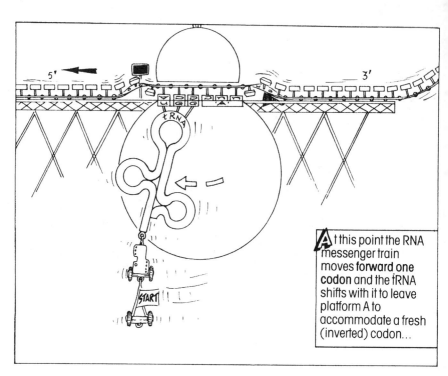

At this point the RNA messenger train moves **forward one codon** and the tRNA shifts with it to leave platform A to accommodate a fresh (inverted) codon...

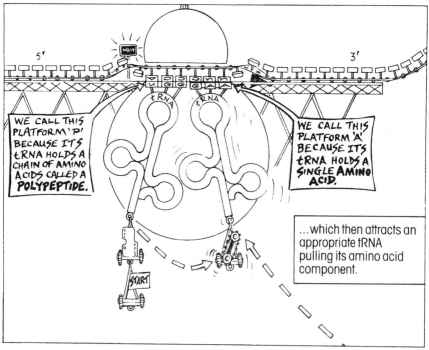

WE CALL THIS PLATFORM 'P' BECAUSE ITS tRNA HOLDS A CHAIN OF AMINO ACIDS CALLED A **POLYPEPTIDE.**

WE CALL THIS PLATFORM 'A' BECAUSE ITS tRNA HOLDS A SINGLE **AMINO ACID.**

...which then attracts an appropriate tRNA pulling its amino acid component.

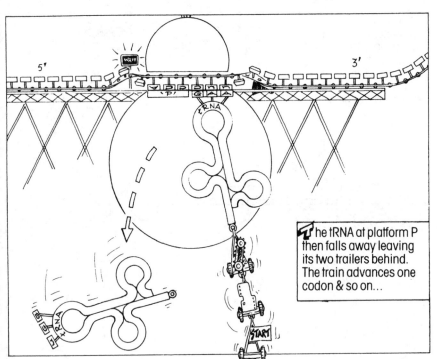

The tRNA at platform P then falls away leaving its two trailers behind. The train advances one codon & so on...

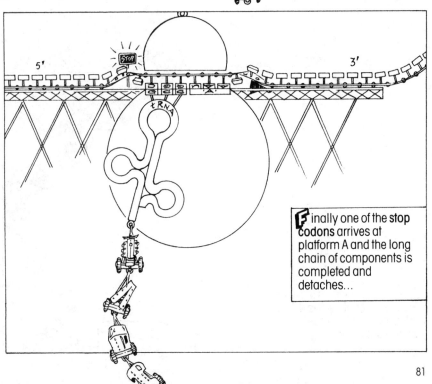

Finally one of the **stop codons** arrives at platform A and the long chain of components is completed and detaches...

...and leaves the ribosome. The automatic ramp drops down, the small unit of the factory rises and returns to its original position as the tail end of the train leaves.

RIBOSOME INC.

PROTEIN TRANSLATING
FACTORY

As the components leave the factory the chain starts to twist & folds together in a most remarkable way...

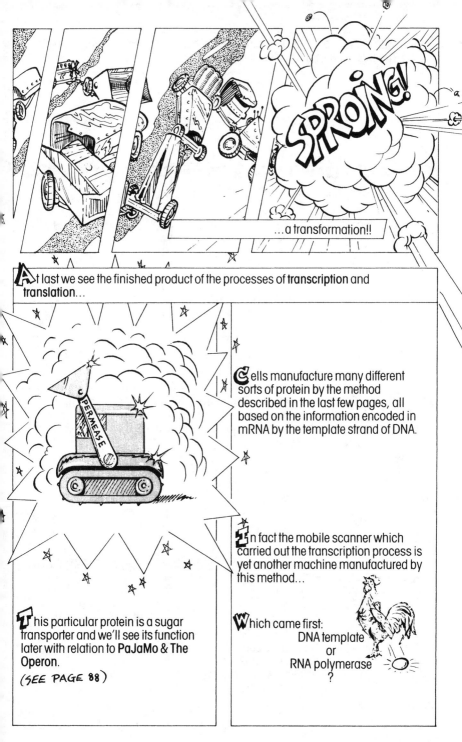

...a transformation!!

At last we see the finished product of the processes of **transcription** and **translation**...

Cells manufacture many different sorts of protein by the method described in the last few pages, all based on the information encoded in mRNA by the template strand of DNA.

In fact the mobile scanner which carried out the transcription process is yet another machine manufactured by this method...

Which came first:
DNA template
or
RNA polymerase
?

This particular protein is a sugar transporter and we'll see its function later with relation to **PaJaMo & The Operon**.

(SEE PAGE 88)

It's worth pointing out that there can be a number of ribosomes all translating the same RNA messenger train at different points.

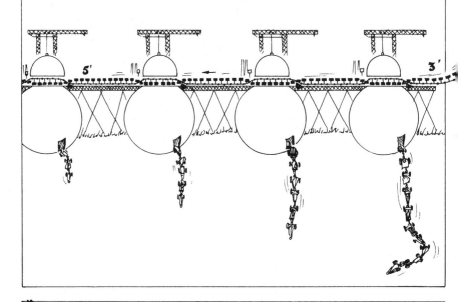

Also that the RNA messenger train is **short-lived** unlike the long-lived template DNA strand from which it was assembled. This means that when the train has passed through the last ribosome, its job having been done...

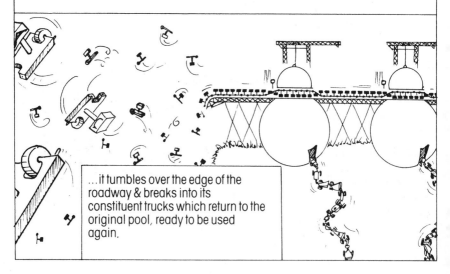

...it tumbles over the edge of the roadway & breaks into its constituent trucks which return to the original pool, ready to be used again.

Genetic Code

The "genetic code" itself was cracked in the early 1960's. **Severo Ochoa** of the New York University Medical School devised enzymatic methods for making RNA molecules in the test tube which had defined nucleotide sequences.

Marshall Nirenberg and his student Phil Leder, working at the National Institutes of Health in Maryland, used synthetic RNA made by Ochoa's methods to direct protein synthesis by cell extracts in the test tube.

Ochoa   Nirenberg

They found that simple RNA trinucleotides, the minimal molecules for specifying a code word, were sufficient for binding tRNA to ribosomes. The RNA triplet would bind to the ribosome, and guide only one tRNA into place. In one typical experiment GUU was the added triplet, and only valine tRNA was bound to the ribosome. Thus GUU is a code word for valine. Remember there are sixty-four code words but only twenty amino acids. Some amino acids can be specified by more than one triplet. Thus ACU, ACC, ACA, and ACG all code for threonine. Only three triplets failed to direct tRNA binding: UAA, UAG, and UGA. These are the "nonsense" or "stop here" signals postulated by Crick and Brenner.

GENETIC CODE CRACKED FULL STORY

1ST → 2ND →	U	C	A	G	3RD
U	PHE PHE LEU LEU	SER SER SER SER	TYR TYR STOP STOP	CYS CYS STOP TRP	U C A G
C	LEU LEU LEU LEU	PRO PRO PRO PRO	HIS HIS GLUN GLUN	ARG ARG ARG ARG	U C A G
A	ILEU ILEU ILEU MET	THR THR THR THR	ASPN ASPN LYS LYS	SER SER ARG ARG	U C A G
G	VAL VAL VAL VAL	ALA ALA ALA ALA	ASP ASP GLU GLU	GLY GLY GLY GLY	U C A G

PHE - PHENYLALANINE
GLU - GLUTAMIC ACID
ASP - ASPARTIC ACID
ASPN - ASPARAGINE
ILEU - ISOLEUCINE
MET - METHIONINE
THR - THREONINE
ARG - ARGININE
GLUN - GLUTAMINE
HIS - HISTIDINE
TRP - TRYPTOPHAN
TYR - TYROSINE
CYS - CYSTEINE
LEU - LEUCINE
PRO - PROLINE
ALA - ALANINE
VAL - VALINE
GLY - GLYCINE
LYS - LYSINE
SER - SERINE

KEY

Here it is. The code for each of the twenty amino
So simple isn't it? Read the table and
can't l

86

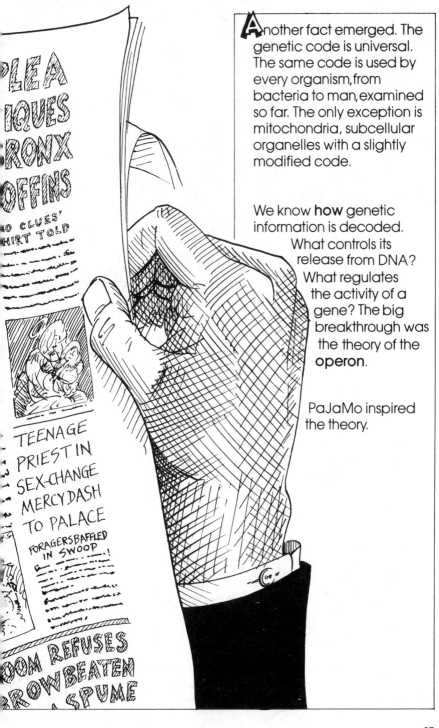

Another fact emerged. The genetic code is universal. The same code is used by every organism, from bacteria to man, examined so far. The only exception is mitochondria, subcellular organelles with a slightly modified code.

We know **how** genetic information is decoded. What controls its release from DNA? What regulates the activity of a gene? The big breakthrough was the theory of the **operon**.

PaJaMo inspired the theory.

PLEA
IQUES
RONX
OFFINS
O CLUES'
HIRT TOLD

TEENAGE
PRIEST IN
SEX-CHANGE
MERCY DASH
TO PALACE
FORAGERS BAFFLED
IN SWOOP

OOM REFUSES
ROWBEATEN
SPUME

87

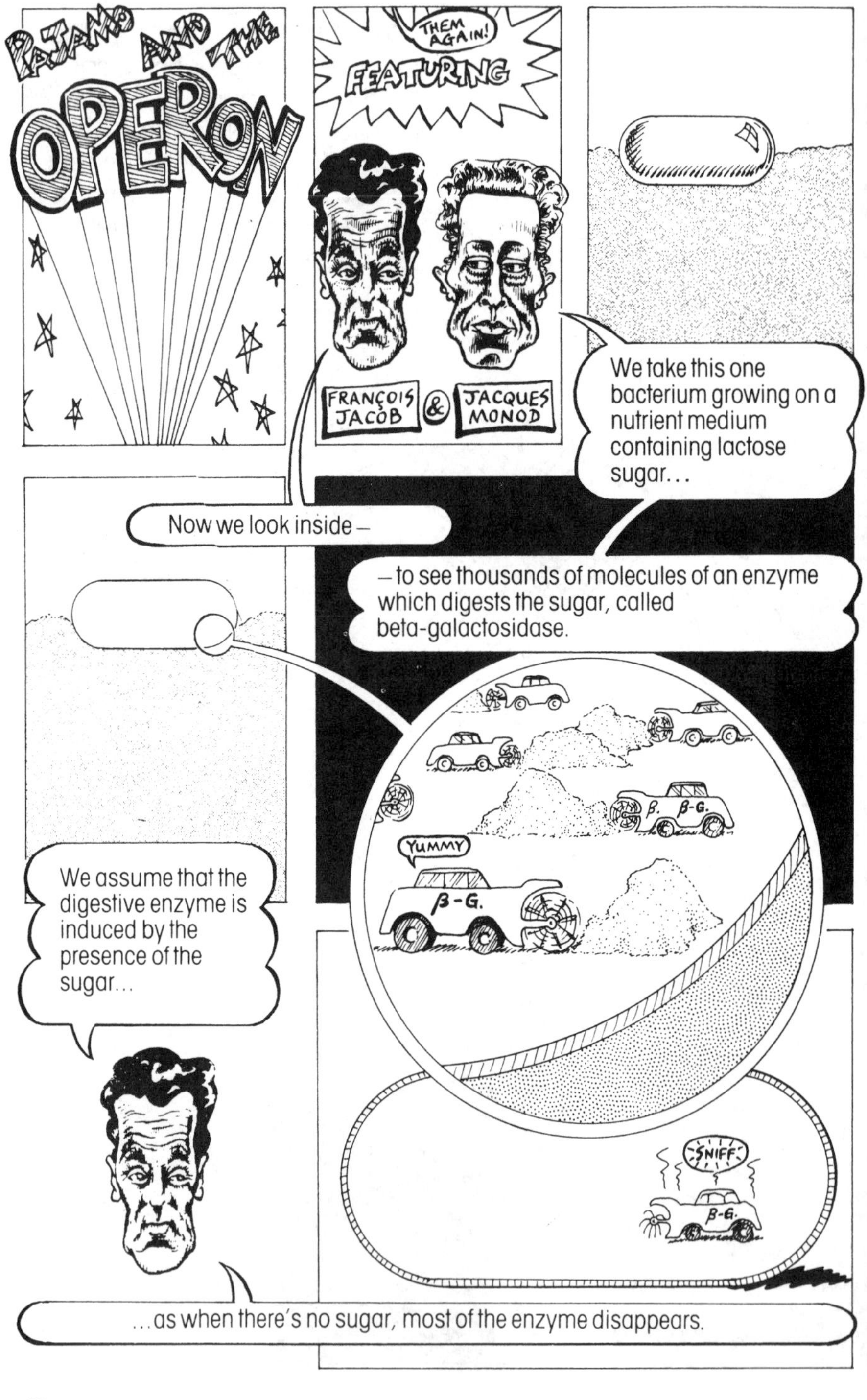
PAJAMO AND THE OPERON
THEM AGAIN!
FEATURING
FRANÇOIS JACOB
&
JACQUES MONOD
We take this one bacterium growing on a nutrient medium containing lactose sugar...
Now we look inside –
– to see thousands of molecules of an enzyme which digests the sugar, called beta-galactosidase.
YUMMY
β-G.
β. β-G.
We assume that the digestive enzyme is induced by the presence of the sugar...
SNIFF
β-G.
...as when there's no sugar, most of the enzyme disappears.

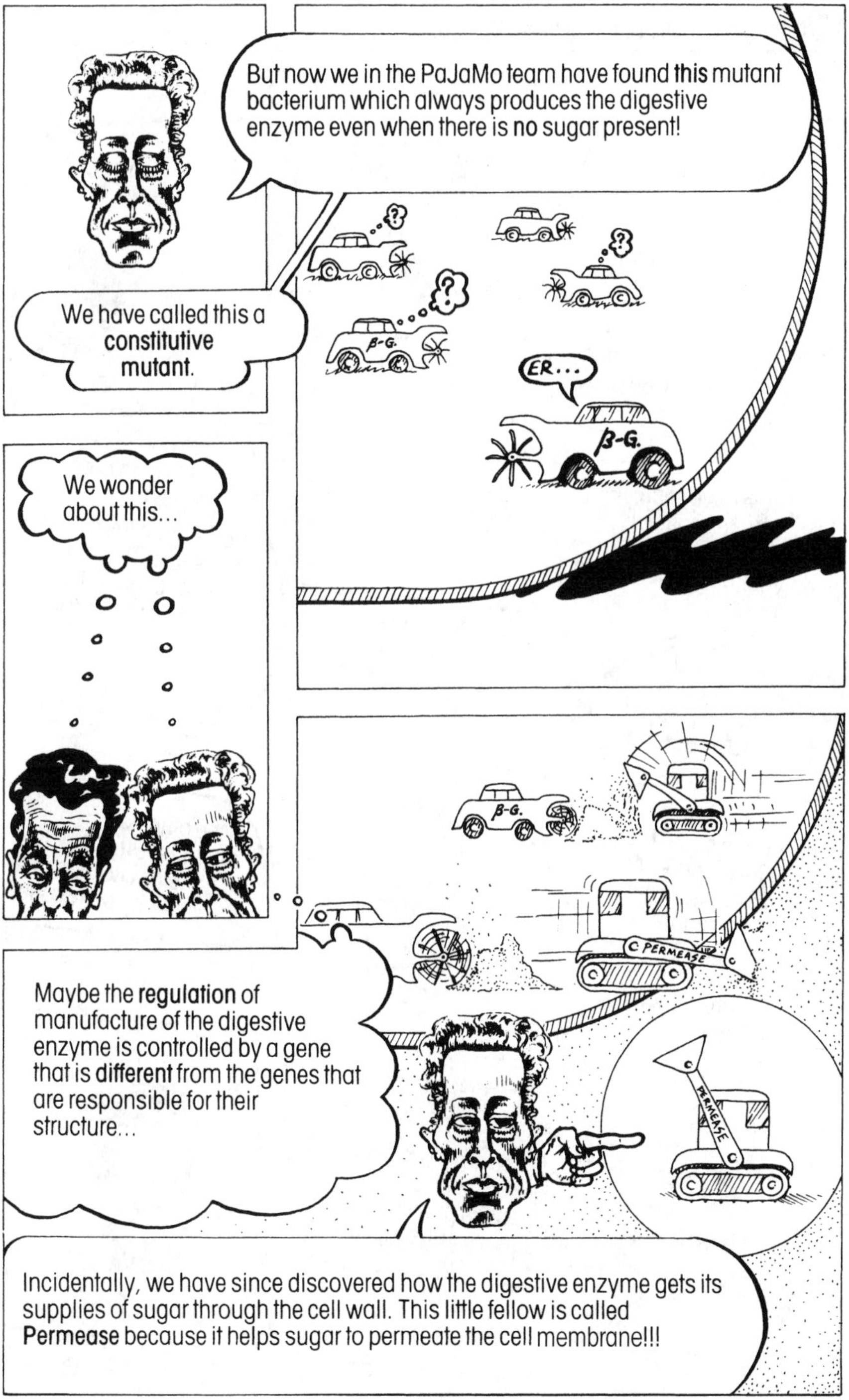

But now we in the PaJaMo team have found this mutant bacterium which always produces the digestive enzyme even when there is no sugar present!
We have called this a constitutive mutant.
β-G.
ER...
β-G.
We wonder about this...
β-G.
PERMEASE
Maybe the regulation of manufacture of the digestive enzyme is controlled by a gene that is different from the genes that are responsible for their structure...
PERMEASE
Incidentally, we have since discovered how the digestive enzyme gets its supplies of sugar through the cell wall. This little fellow is called Permease because it helps sugar to permeate the cell membrane!!!

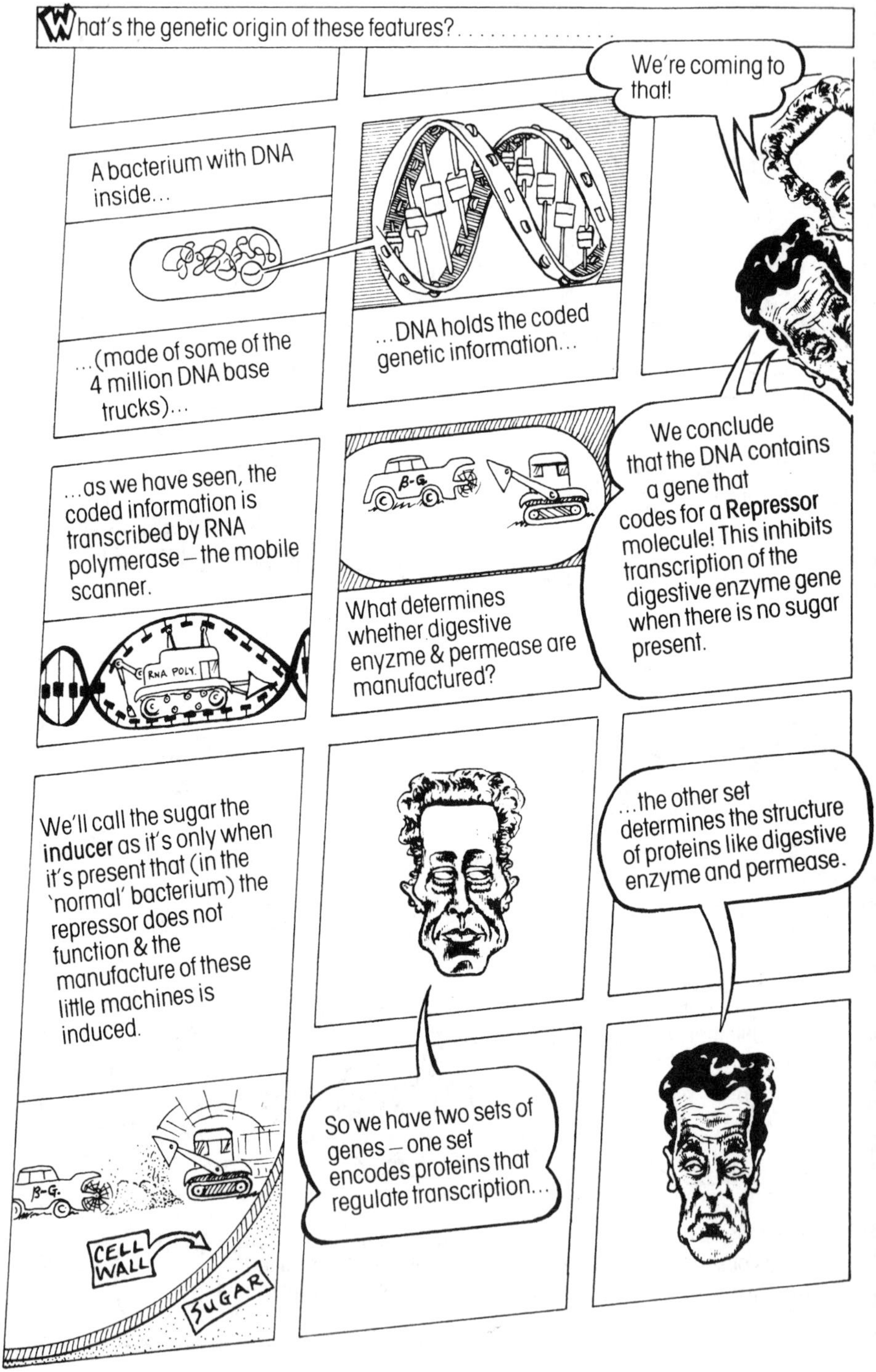
What's the genetic origin of these features?...............
We're coming to that!
A bacterium with DNA inside...
...(made of some of the 4 million DNA base trucks)...
...DNA holds the coded genetic information...
We conclude that the DNA contains a gene that codes for a **Repressor** molecule! This inhibits transcription of the digestive enzyme gene when there is no sugar present.
...as we have seen, the coded information is transcribed by RNA polymerase – the mobile scanner.
RNA POLY.
β-G.
What determines whether digestive enyzme & permease are manufactured?
We'll call the sugar the **inducer** as it's only when it's present that (in the 'normal' bacterium) the repressor does not function & the manufacture of these little machines is induced.
β-G.
CELL WALL
SUGAR
So we have two sets of genes – one set encodes proteins that regulate transcription...
...the other set determines the structure of proteins like digestive enzyme and permease.

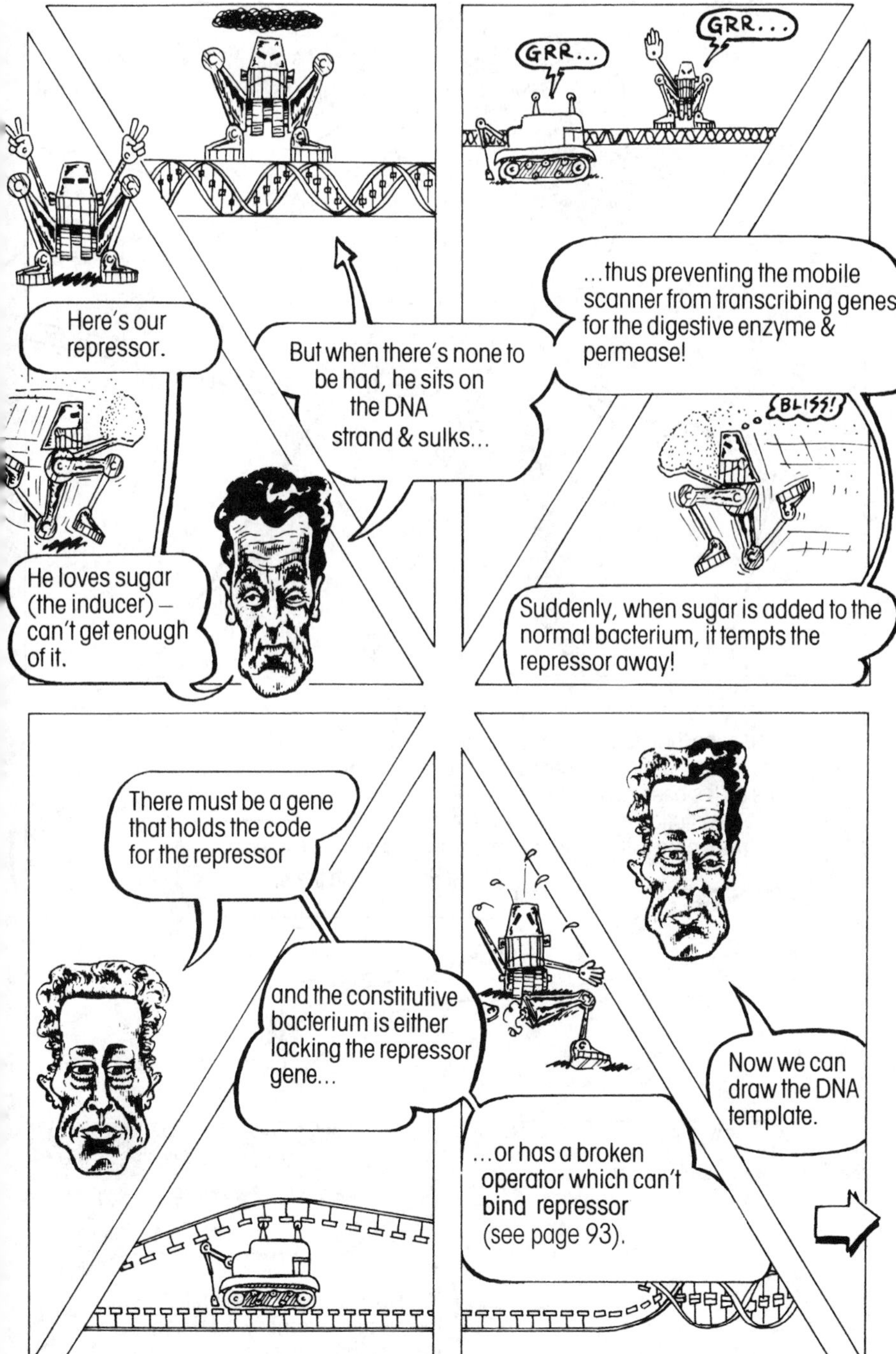

Here's our repressor.
He loves sugar (the inducer) – can't get enough of it.
But when there's none to be had, he sits on the DNA strand & sulks...
GRR...
GRR...
...thus preventing the mobile scanner from transcribing genes for the digestive enzyme & permease!
BLISS!
Suddenly, when sugar is added to the normal bacterium, it tempts the repressor away!
There must be a gene that holds the code for the repressor
and the constitutive bacterium is either lacking the repressor gene...
...or has a broken operator which can't bind repressor (see page 93).
Now we can draw the DNA template.

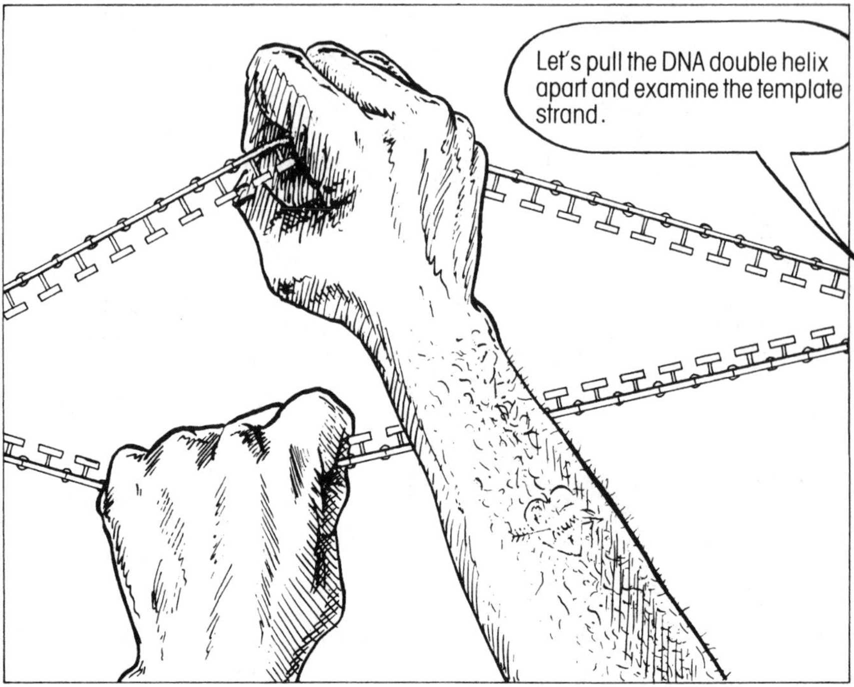
Let's pull the DNA double helix apart and examine the template strand.

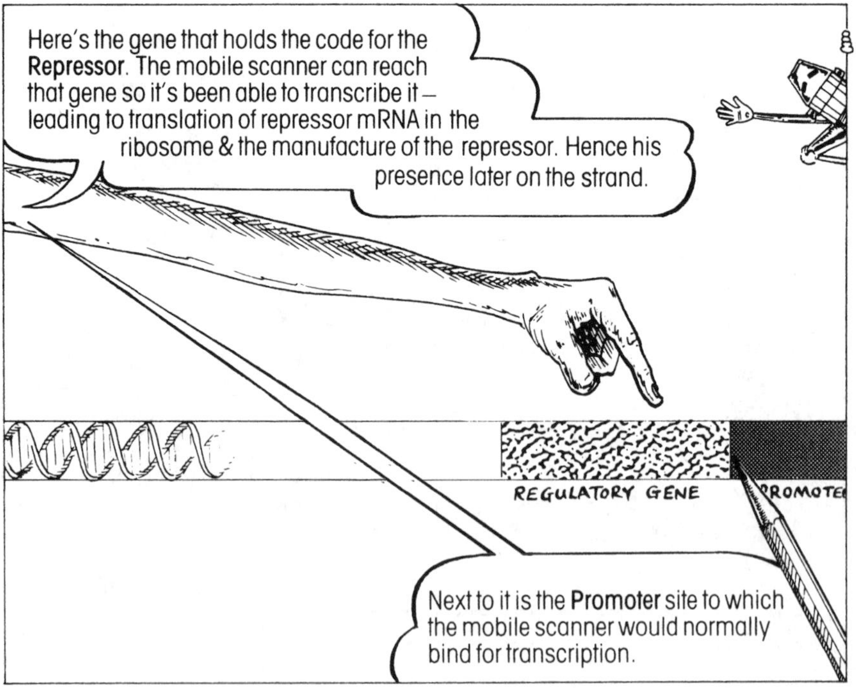
Here's the gene that holds the code for the **Repressor**. The mobile scanner can reach that gene so it's been able to transcribe it – leading to translation of repressor mRNA in the ribosome & the manufacture of the repressor. Hence his presence later on the strand.
REGULATORY GENE
Next to it is the **Promoter** site to which the mobile scanner would normally bind for transcription.

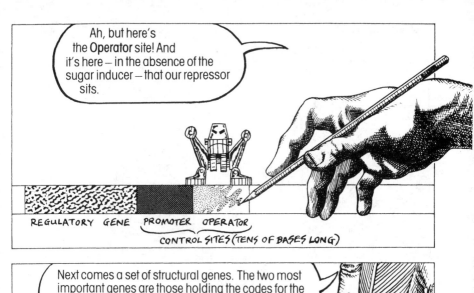

Ah, but here's the **Operator** site! And it's here — in the absence of the sugar inducer — that our repressor sits.

REGULATORY GENE   PROMOTER   OPERATOR

CONTROL SITES (TENS OF BASES LONG)

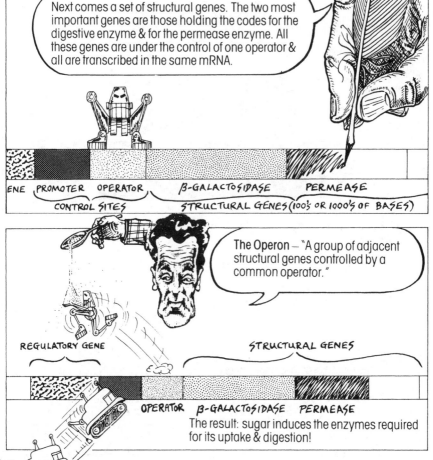

Next comes a set of structural genes. The two most important genes are those holding the codes for the digestive enzyme & for the permease enzyme. All these genes are under the control of one operator & all are transcribed in the same mRNA.

ENE   PROMOTER   OPERATOR   β-GALACTOSIDASE   PERMEASE

CONTROL SITES   STRUCTURAL GENES (100's OR 1000's OF BASES)

The Operon — "A group of adjacent structural genes controlled by a common operator."

REGULATORY GENE   STRUCTURAL GENES

OPERATOR   β-GALACTOSIDASE   PERMEASE

The result: sugar induces the enzymes required for its uptake & digestion!

93

he Operon model disclosed by PaJaMo ranks with
Crick's Adaptor Hypothesis...

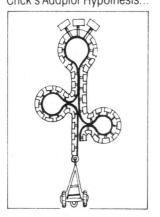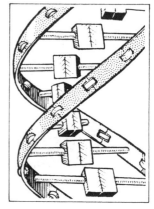

...and the Watson-Crick model as one
of the major intellectual achievements of modern biology!

AW, SHUCKS

IT WAS NUTHIN'

Operon control was only the first form of bacterial gene regulation to be discovered!

Since the 1960's many others, just as fascinating, have been found!

he operon provides diversity of gene expression for the **individual organism**, in response to hour by hour changes in the environment. But what created the diversity of the genes **themselves** that reside in different organisms? This is the diversity of variation and of speciation itself.

THE DIVERSITY OF GENE EXPRESSION

Mutation is one source of gene diversity.

But life on earth began some 3 to 4 billion years ago.

I DON'T SEEM TO BE GETTING VERY FAR

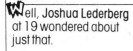

If the only cause of variation was random mutation, evolution would have been very slow!

Sexual reproduction could have provided great variability in primitive organisms by reshuffling mutations. But can primitive organisms like bacteria have sex?

Bacterial sex seems absurd!

I'm puzzled by Avery's results.

Well, Joshua Lederberg at 19 wondered about just that.

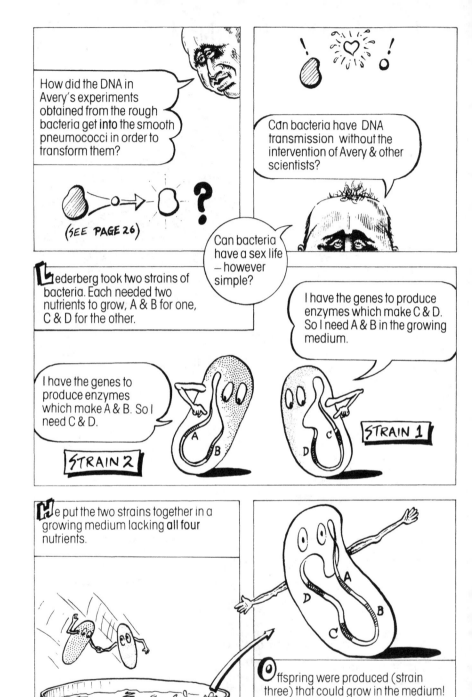

How did the DNA in Avery's experiments obtained from the rough bacteria get **into** the smooth pneumococci in order to transform them?

(SEE PAGE 26)

Can bacteria have DNA transmission without the intervention of Avery & other scientists?

Can bacteria have a sex life – however simple?

ederberg took two strains of bacteria. Each needed two nutrients to grow, A & B for one, C & D for the other.

I have the genes to produce enzymes which make C & D. So I need A & B in the growing medium.

I have the genes to produce enzymes which make A & B. So I need C & D.

STRAIN 2

STRAIN 1

e put the two strains together in a growing medium lacking **all four** nutrients.

ffspring were produced (strain three) that could grow in the medium!

Because mutations are rare (one in a million) & because I am using bacterial strains 1 & 2, needing **two** nutrient substances each, the chances of getting bacteria which need none of these nutrients through mutation are practically zero (one million times one million)!

Strain 3 bacteria must have been produced through sexual conjugation (that is, transmission of DNA) between strains 1 & 2.

Bacteria have a sex life!

97

Having discovered that bacteria have a sex life, biologists soon found that they have **sexes** as well, and explained how genes were transferred from **1** to **2** to make **3**. Male bacteria (or what biologists have dubbed the males) have a piece of DNA called the **F-factor** (fertility factor). Those bacteria that have the F-factor have special appendages called **sex-pili** which bind to receptor sites on the female bacterium (bacteria without the F-factor). A DNA molecule is transferred from the male to the female through the sex-pili!

DNA from the male enters the female in a linear fashion. (Jacob called this the spaghetti hypothesis.) If A, B, C, D, E, F, G, ... Z are the genes in the male bacterium, first A, then B, etc. will enter the bacteria in order and at a fixed rate of entry. They might enter starting anywhere in the sequence (say J), then move on to Z and continue through A to I. In this process, the F-factor (DNA with genes for "maleness") may leave the male and enter the female, making the female a male and the former male a female! Maleness, as some biologists liked to say, is contagious!

Other pieces of DNA can enter a bacterium. One, a small DNA circle which remains separate from the bacterial chromosome, is called a plasmid. It's very useful in the cloning of genes, as we'll soon see.

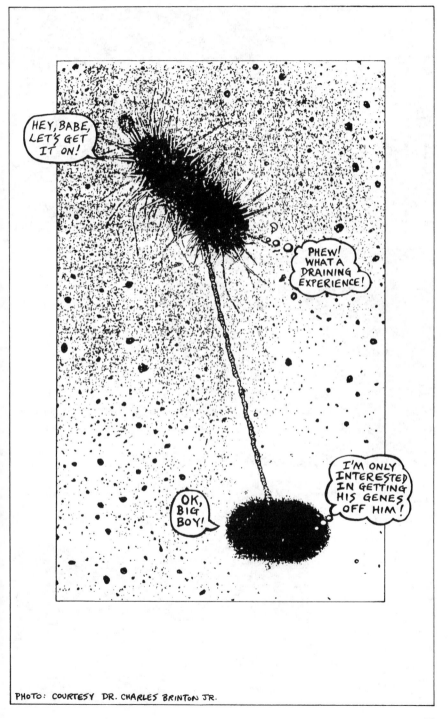

PHOTO: COURTESY DR. CHARLES BRINTON JR.

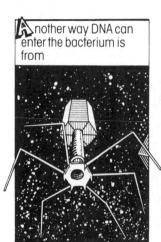

Another way DNA can enter the bacterium is from

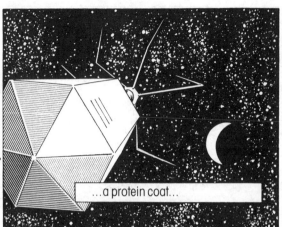

...a protein coat...

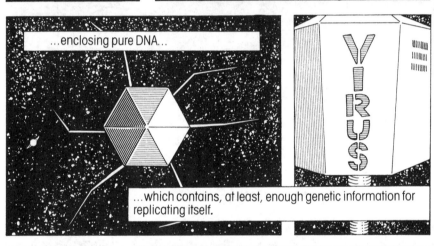

...enclosing pure DNA...

...which contains, at least, enough genetic information for replicating itself.

VIRUS

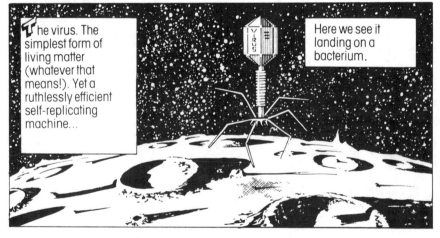

The virus. The simplest form of living matter (whatever that means!). Yet a ruthlessly efficient self-replicating machine...

Here we see it landing on a bacterium.

n 1952, **Martha Chase & Alfred Hershey** performed an experiment using a food blender which showed that the genetic information of the virus was contained in its DNA, not its protein.

(Scientists call such a virus a bacteriophage — or **phage** for short.)

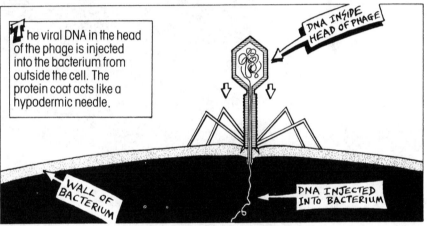

he viral DNA in the head of the phage is injected into the bacterium from outside the cell. The protein coat acts like a hypodermic needle.

DNA INSIDE HEAD OF PHAGE

WALL OF BACTERIUM

DNA INJECTED INTO BACTERIUM

Once the viral DNA is inside it can take control of the bacterial cell and by expressing its genes using the bacterium's mobile scanners, tRNA's & ribosomes, force the bacterium to produce hundreds of new viral coats & hundreds of copies of viral DNA.

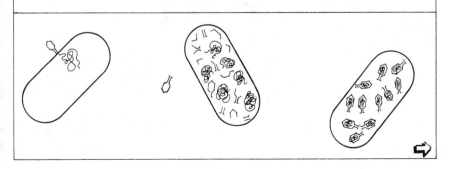

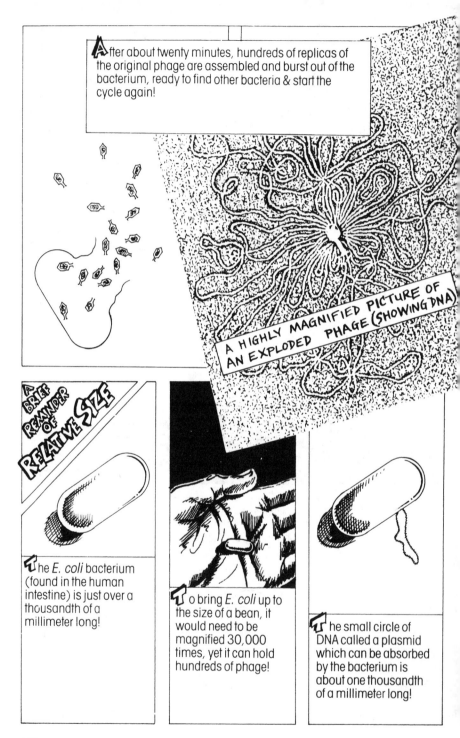

After about twenty minutes, hundreds of replicas of the original phage are assembled and burst out of the bacterium, ready to find other bacteria & start the cycle again!

A HIGHLY MAGNIFIED PICTURE OF AN EXPLODED PHAGE (SHOWING DNA)

A BRIEF REMINDER OF RELATIVE SIZE

The *E. coli* bacterium (found in the human intestine) is just over a thousandth of a millimeter long!

To bring *E. coli* up to the size of a bean, it would need to be magnified 30,000 times, yet it can hold hundreds of phage!

The small circle of DNA called a plasmid which can be absorbed by the bacterium is about one thousandth of a millimeter long!

All these complex processes going on in a speck invisible to the eye!! No wonder the beginner is as baffled by the impenetrable "inner space" of the cell as by the infinite mysteries of the universe!!

Flabbergasted? Let's further confound you with a glimpse of the complex inner world of the **cell**. A "typical" cell from a mammal contains enzymes, other proteins, fats, sugars, amino acids, other building blocks & energy — carrying molecules **plus** a yard of double-stranded DNA! The double helix is a little less than 80 billionths of an inch in diameter! A single full twist in the molecule measures just 134 billionths of an inch!

Phew! Having looked at the many different contributors to classical molecular biology, & their discoveries, I think it's time to move into modern genetics, recent research & pointers to the future…

103

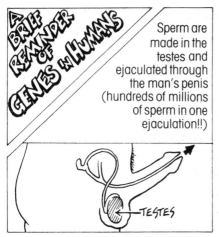

A BRIEF REMINDER OF GENES IN HUMANS

Sperm are made in the testes and ejaculated through the man's penis (hundreds of millions of sperm in one ejaculation!!)

TESTES

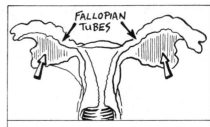

FALLOPIAN TUBES

Eggs are stored in the woman's ovaries & released at the rate of one every four weeks; they lodge in the fallopian tubes awaiting fertilization by one of the sperm.

A single sperm cell (at the head of the lashing tail) is mainly made up of **nucleus**.

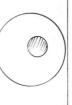

The egg cell (smaller than a pinhead) has jelly-like **cytoplasm** enclosing a nucleus.

Inside the nucleus is a darker-staining material known as **chromatin** made up of fine tangled threads.

Now, when a cell is about to divide the chromatin contracts into groups called chromosomes. A human has 46 chromosomes.

The sperm nucleus has 23 chromosomes (as has the egg nucleus). When sperm fertilizes egg, their nuclei fuse and the cell has 46 chromosomes once again.

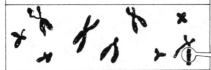

The chromosomes group into 23 **pairs** each with one member from sperm & the other from egg. (SEE PAGE 154)

To bring one of the larger paired chromosomes up to a centimeter in length it needs to be magnified about 2000 times! Chromosomes are composed of the tiny threads of DNA. (MORE ON CHROMOSOMES ON PAGE 154)

Prokaryotes versus Eukaryotes

With the genetic code cracked, and the outlines of genetic regulation firmly established for bacteria and their viruses, scientists began to confront the awesome problem of gene structure and regulation in higher eukaryotes (those having cells **with** nuclei) including man.

Some dismissed the problem altogether saying:

> The findings are likely to be mere recapitulations of the rules already discovered in bacteria.

Others argued that:

> Fundamental differences between bacteria (which are prokaryotes — cells without nuclei) prevent them from using the very same mechanisms for gene regulation & expression as animal & plant cells (which are eukaryotes & have nuclei).

In **eukaryotes**, because the DNA is contained in the nucleus, it is in a compartment isolated from the translation machinery, which resides in the cytoplasm.

Bacteria, because they lack nuclei, carry out transcription and translation side by side.

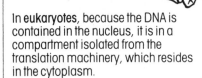

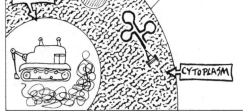

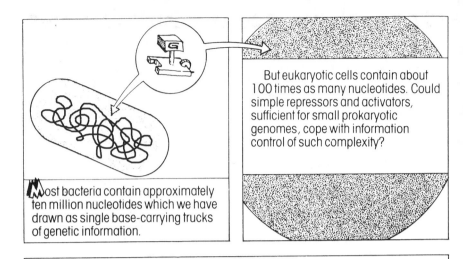

Most bacteria contain approximately ten million nucleotides which we have drawn as single base-carrying trucks of genetic information.

But eukaryotic cells contain about 100 times as many nucleotides. Could simple repressors and activators, sufficient for small prokaryotic genomes, cope with information control of such complexity?

Genetics would be one way to study animal cell gene expression, but genetic studies of eukaryotic gene expression were cumbersome. Some of these problems were solved, however, by animal cell tissue culture. In tissue culture, cells are isolated from an animal, and are propagated apart from the animal, in the laboratory, in specially concocted culture media. Unlike bacteria, animal cells are not equipped for infinite division, for in normal circumstances they perish with the death of the individual. After much trial and error, "immortalized" cells, capable of indefinite growth in the laboratory, were established.

One human cell line, HeLa, has been propagated in the laboratory for over sixty years. Cultured cells provide the convenience of bacteria, and permit experiments impossible with the whole animal.

With tissue cultured cells, a form of sexuality may be achieved through "cell fusion."

If different cultured cell types are mingled together in the presence of certain viruses or chemical agents, they clump and merge their exterior membranes. The nuclei of the clumped cells now occupy a single fused cell.

During cell division, the nuclear membranes of the original cells disintegrate, and the sets of chromosomes comingle. Cell types from the different species, such as human and mouse cells, can be conveniently fused, generating a hybrid man-mouse cell line.

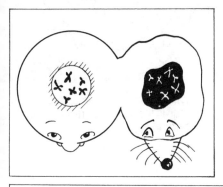

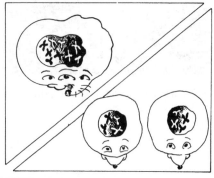

Superficially, cell fusion resembles the comingling of sperm and egg chromosomes after fertilization.

Cells with both human and mouse chromosomes may be obtained in the lab, although they never develop into a hybrid multi-cellular organism such as a mouse-man!

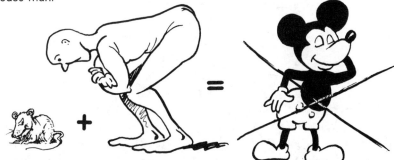

Unfortunately, cells in tissue culture often lose specialized differentiated properties, and their suitability as models for differentiated gene expression is in question.

The problems of **differentiation** raise another major difference between animal cells and bacteria.

A soil bacterium leads a lonely and difficult existence.

It must, from time to time, adapt to change in the nutrients it receives from its surroundings.

A bacterium can undergo many cell divisions without altering the repertoire of gene expression responses it can muster.

Animal cells reside (in general) within the organism, bathed in an unchanging environment of body fluids and tissues. They need not respond to dramatic changes in the environment. Their gene expression programs are geared to doing specific jobs in the organism (differentiation).

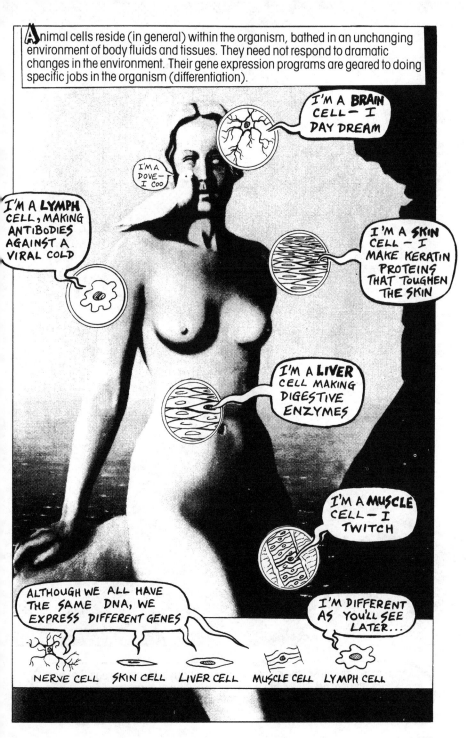

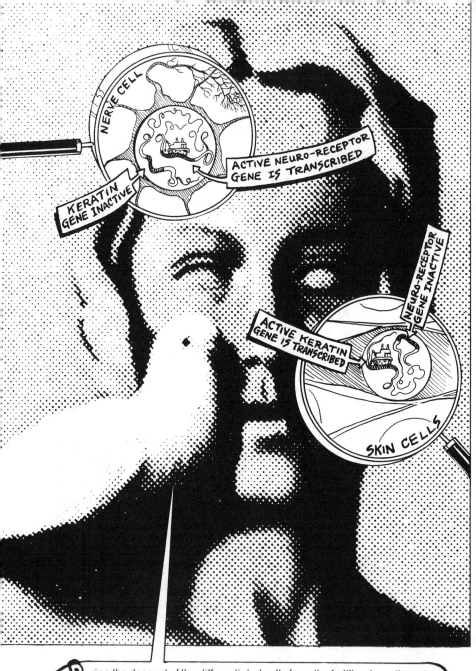

During the descent of the differentiated cells from the fertilized egg the very same complement of DNA is retained in each cell (with few exceptions). Differentiation does not result from shedding unwanted genes.

hus, both bacteria and differentiated cells maintain a constant DNA content during gene regulation.

But a bacterium may switch a gene on and off virtually an infinite number of times.

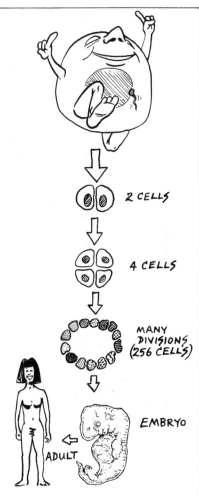

2 CELLS

4 CELLS

MANY DIVISIONS (256 CELLS)

EMBRYO

ADULT

In contrast, a eukaryotic cell, once differentiated, generally does not change its expression to become another type; that is, brain cells cannot become liver cells or vice versa. (But for a laboratory procedure that can change cell types, see page 176.)

The bacterium copes in a solitary and repetitive manner with fluctuations in its environment (as the bacterium in the PaJaMo experiment reacts to the absence or presence of sugar).

The eukaryotic fertilized egg throws caution to the wind. It gives rise to a wide range of differentiated cells that form the organism and eventually die. Unlike the bacterium, it can employ regulation mechanisms which are, practically speaking, irreversible.

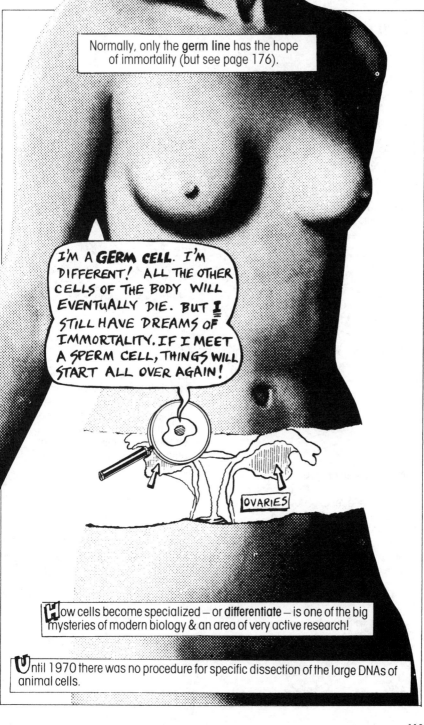

Normally, only the **germ line** has the hope of immortality (but see page 176).

I'M A **GERM CELL**. I'M DIFFERENT! ALL THE OTHER CELLS OF THE BODY WILL EVENTUALLY DIE. BUT I STILL HAVE DREAMS OF IMMORTALITY. IF I MEET A SPERM CELL, THINGS WILL START ALL OVER AGAIN!

OVARIES

How cells become specialized — or **differentiate** — is one of the big mysteries of modern biology & an area of very active research!

Until 1970 there was no procedure for specific dissection of the large DNAs of animal cells.

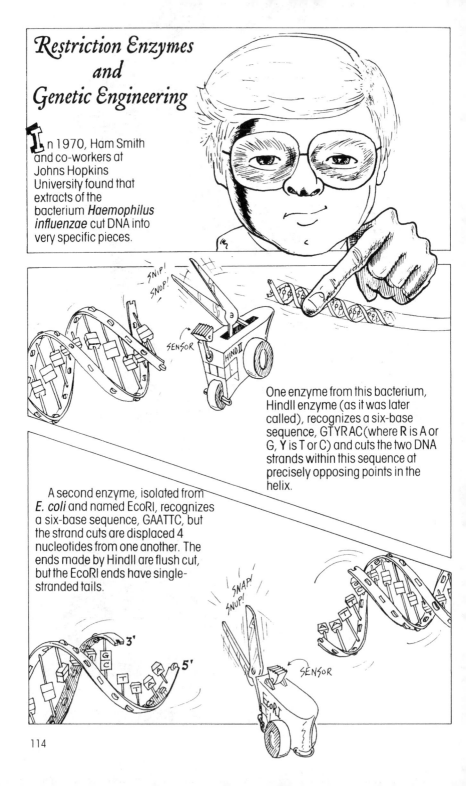

Restriction Enzymes and Genetic Engineering

In 1970, Ham Smith and co-workers at Johns Hopkins University found that extracts of the bacterium *Haemophilus influenzae* cut DNA into very specific pieces.

One enzyme from this bacterium, HindII enzyme (as it was later called), recognizes a six-base sequence, GTYRAC (where R is A or G, Y is T or C) and cuts the two DNA strands within this sequence at precisely opposing points in the helix.

A second enzyme, isolated from *E. coli* and named EcoRI, recognizes a six-base sequence, GAATTC, but the strand cuts are displaced 4 nucleotides from one another. The ends made by HindII are flush cut, but the EcoRI ends have single-stranded tails.

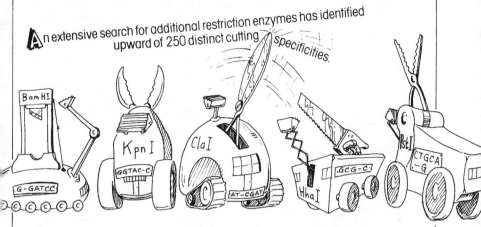

An extensive search for additional restriction enzymes has identified upward of 250 distinct cutting specificities.

BamHI
G-GATCC

KpnI
GGTAC-C

ClaI
AT-CGAT

HhaI
·GCG·C·

PstI
CTGCA
-G

The enzymes are present in a wide range of single-celled organisms: in many bacteria, and in yeast as well. Restriction enzymes provide an invaluable tool for dissecting complex DNA genomes at specific points.

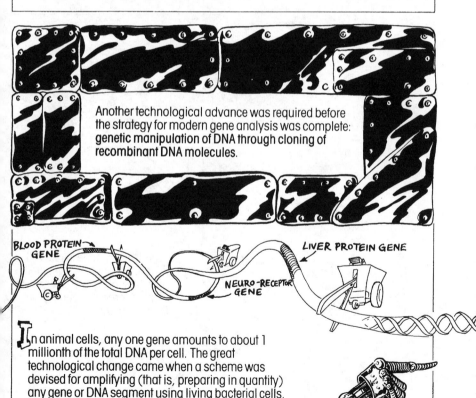

Another technological advance was required before the strategy for modern gene analysis was complete: **genetic manipulation of DNA through cloning of recombinant DNA molecules.**

BLOOD PROTEIN GENE

NEURO-RECEPTOR GENE

LIVER PROTEIN GENE

In animal cells, any one gene amounts to about 1 millionth of the total DNA per cell. The great technological change came when a scheme was devised for amplifying (that is, preparing in quantity) any gene or DNA segment using living bacterial cells.

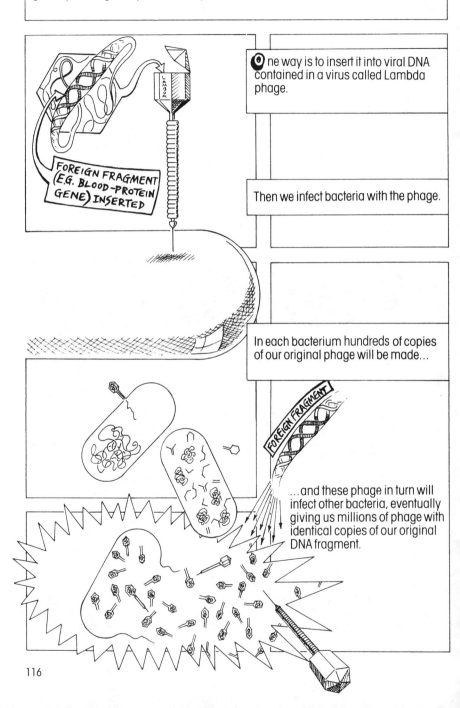

We now know how to dissect a gene. But how can we make use of a natural replicating system to amplify a gene — that is, increase the amount of a specific gene by making many identical copies of it?

One way is to insert it into viral DNA contained in a virus called Lambda phage.

FOREIGN FRAGMENT (E.G. BLOOD-PROTEIN GENE) INSERTED

Then we infect bacteria with the phage.

In each bacterium hundreds of copies of our original phage will be made...

FOREIGN FRAGMENT

...and these phage in turn will infect other bacteria, eventually giving us millions of phage with identical copies of our original DNA fragment.

WE CAN INSERT OUR DNA FRAGMENT INTO A PLASMID (A SMALL CIRCULAR PIECE OF DNA THAT CAN REPLICATE IN A BACTERIUM) . . .

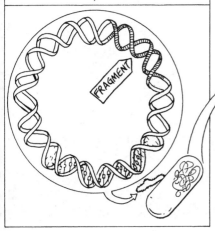

FRAGMENT

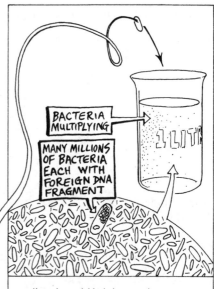

BACTERIA MULTIPLYING

MANY MILLIONS OF BACTERIA EACH WITH FOREIGN DNA FRAGMENT

1 LITRE

...the plasmid is taken up by a bacterium & replicates alongside the bacterial DNA.

While phage rely on special structures to enter cells, plasmids enter by the inefficient and arduous route of traversing the bacterial membrane as a naked DNA molecule. This closely resembles the entry of transforming principle into bacteria in the Griffith-Avery experiments.

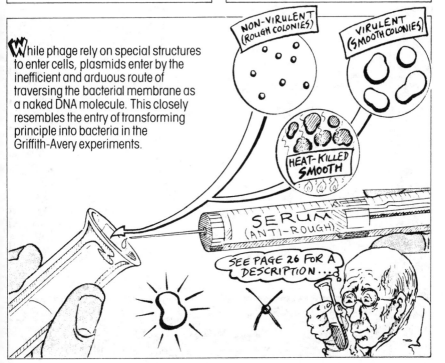

NON-VIRULENT (ROUGH COLONIES)

VIRULENT (SMOOTH COLONIES)

HEAT-KILLED SMOOTH

SERUM (ANTI-ROUGH)

SEE PAGE 26 FOR A DESCRIPTION . . .

117

How do we join our foreign gene to the plasmid DNA?

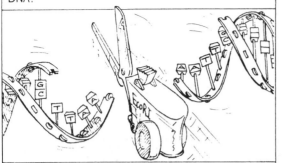

Recall that the EcoRI enzyme makes a staggered break in the double helix.

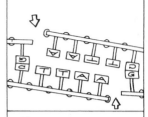

Short, single-stranded tails, four bases long, remain at the cut ends. EcoRI always leaves the tails: AATT.

When a circular DNA molecule (such as a plasmid DNA) which has a single cleavage site is treated with such an enzyme, the circle opens out!

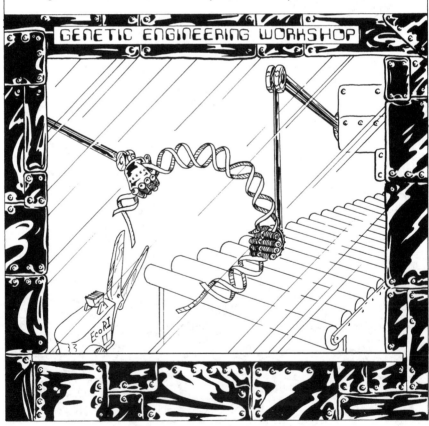

GENETIC ENGINEERING WORKSHOP

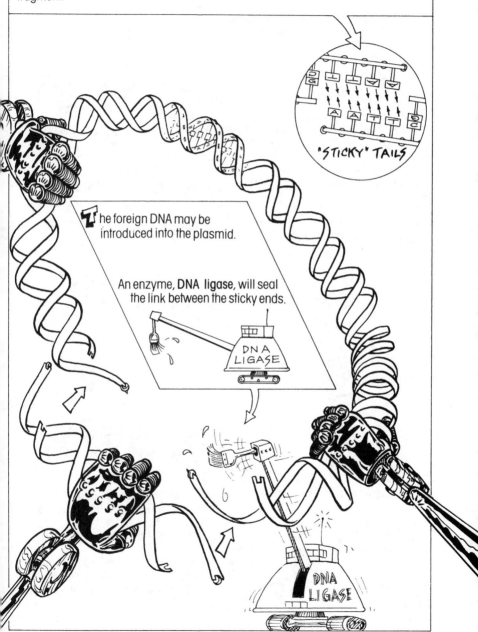

Under certain conditions, the circle can reclose when the single-stranded tails bind together again. The base-pairing of one tail to the other provides the binding force. The tails are "sticky." Tails of plasmid will stick to the tails of a foreign fragment.

"STICKY" TAILS

The foreign DNA may be introduced into the plasmid.

An enzyme, **DNA ligase**, will seal the link between the sticky ends.

DNA LIGASE

DNA LIGASE

119

If we transfer a plasmid vector
(several thousand base-pairs)

with foreign DNA thus inserted (recombinant DNA)

to a bacterium having a chromosome
(about four million base pairs),

we obtain a strain harboring our foreign DNA fragment in a form which will be
amplified during bacterial growth.

Now there are many ways to join DNA fragments and produce them in quantity.

Phage can be manipulated in a
similar way.

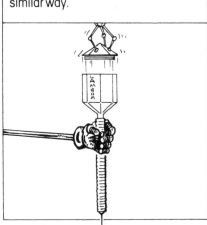

We can insert **foreign DNA** into viral
DNA.

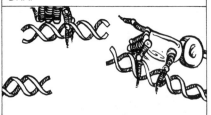

The resulting recombinant DNA can
then be incubated with "**packaging
extracts**" that reassemble our DNA into
biologically active viral structures!

These can infect a large culture of
bacteria, make vast quantities of virus,
& many copies of our foreign fragment.

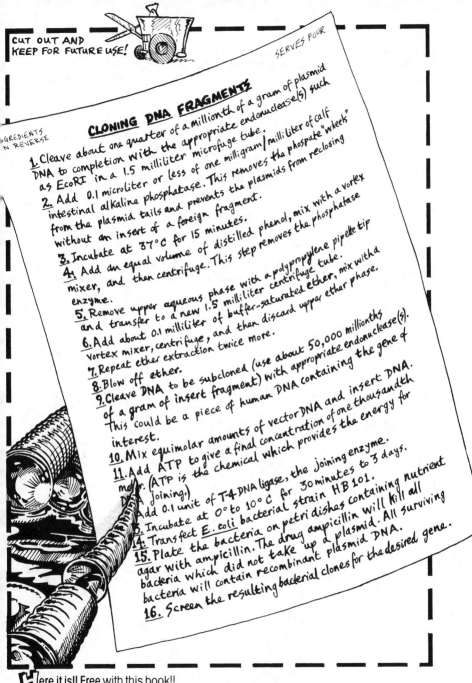

SERVES FOUR

INGREDIENTS
IN REVERSE

CLONING DNA FRAGMENTS

1. Cleave about one quarter of a millionth of a gram of plasmid DNA to completion with the appropriate endonuclease(s) such as EcoRI in a 1.5 milliliter microfuge tube.

2. Add 0.1 microliter or less of one milligram/milliliter of calf intestinal alkaline phosphatase. This removes the phosphate "wheels" from the plasmid tails and prevents the plasmids from reclosing without an insert of a foreign fragment.

3. Incubate at 37°C for 15 minutes.

4. Add an equal volume of distilled phenol, mix with a vortex mixer, and then centrifuge. This step removes the phosphatase enzyme.

5. Remove upper aqueous phase with a polypropylene pipette tip and transfer to a new 1.5 milliliter centrifuge tube.

6. Add about 0.1 milliliter of buffer-saturated ether, mix with a vortex mixer, centrifuge, and then discard upper ether phase.

7. Repeat ether extraction twice more.

8. Blow off ether.

9. Cleave DNA to be subcloned (use about 50,000 millionths of a gram of insert fragment) with appropriate endonuclease(s). This could be a piece of human DNA containing the gene of interest.

10. Mix equimolar amounts of vector DNA and insert DNA.

11. Add ATP to give a final concentration of one thousandth molar. (ATP is the chemical which provides the energy for DNA joining.)

12. Add 0.1 unit of T-4 DNA ligase, the joining enzyme.

13. Incubate at 0° to 10°C for 30 minutes to 3 days.

14. Transfect E. coli bacterial strain HB 101.

15. Plate the bacteria on petri dishes containing nutrient agar with ampicillin. The drug ampicillin will kill all bacteria which did not take up a plasmid. All surviving bacteria will contain recombinant plasmid DNA.

16. Screen the resulting bacterial clones for the desired gene.

Here it is!! Free with this book!!

Your very own recipe for cloning DNA fragments (with explanatory notes).

Sometimes the **foreign DNA** is a pure, well-characterized fragment. However, often we must insert a **mixture** of fragments. This will be the case, for example, if the fragments for insertion were produced by restriction enzyme cleavage of **whole human DNA**.

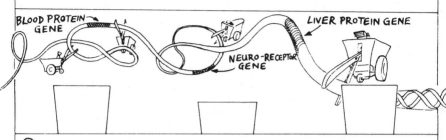

BLOOD PROTEIN GENE

LIVER PROTEIN GENE

NEURO-RECEPTOR GENE

Our purpose might be to amplify & identify a specific human gene. In this case we will have to isolate it from the tens of thousands of other human genes present in human DNA.

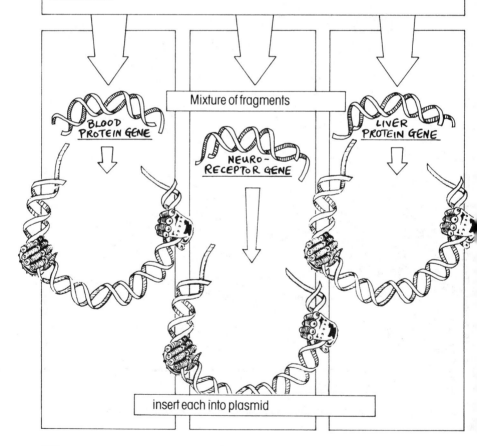

Mixture of fragments

BLOOD PROTEIN GENE

NEURO-RECEPTOR GENE

LIVER PROTEIN GENE

insert each into plasmid

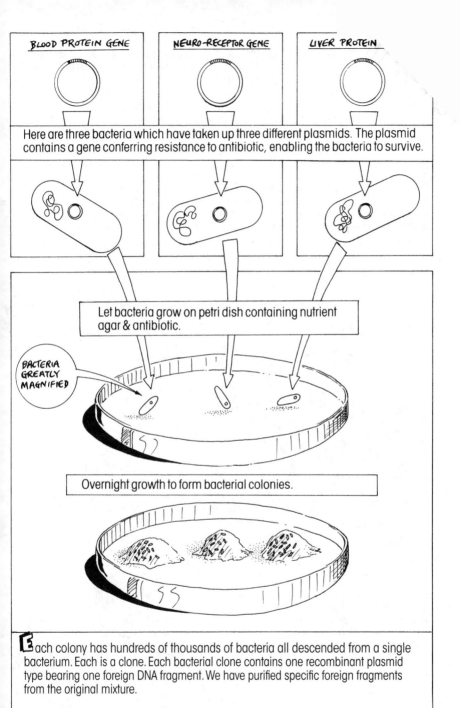

BLOOD PROTEIN GENE   NEURO-RECEPTOR GENE   LIVER PROTEIN

Here are three bacteria which have taken up three different plasmids. The plasmid contains a gene conferring resistance to antibiotic, enabling the bacteria to survive.

Let bacteria grow on petri dish containing nutrient agar & antibiotic.

BACTERIA GREATLY MAGNIFIED

Overnight growth to form bacterial colonies.

Each colony has hundreds of thousands of bacteria all descended from a single bacterium. Each is a clone. Each bacterial clone contains one recombinant plasmid type bearing one foreign DNA fragment. We have purified specific foreign fragments from the original mixture.

All of the plasmids in one bacterial colony descend from a single "parent" plasmid — the one which originally entered the bacterium. They are all identical, and constitute a "clone." A foreign DNA fragment amplified by insertion [in] a plasmid in this manner is said to be "cloned." Cloning therefore [can] provide large quantities of a pure gene which normally exists only in minute quantities in the cell.

Now that any gene may be made plentiful through cloning, how shall we study it? Most of the information content of a gene lies in the precise sequence of the nucleotides. Therefore, it was natural to attempt to determine the precise structures of genes by analyzing their sequences.

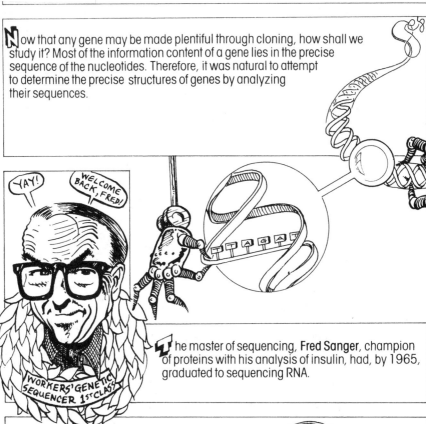

The master of sequencing, **Fred Sanger**, champion of proteins with his analysis of insulin, had, by 1965, graduated to sequencing RNA.

With the challenge of DNA looming before him, Sanger devised a technique for DNA sequencing which used cloning technology and DNA synthesis enzymology.

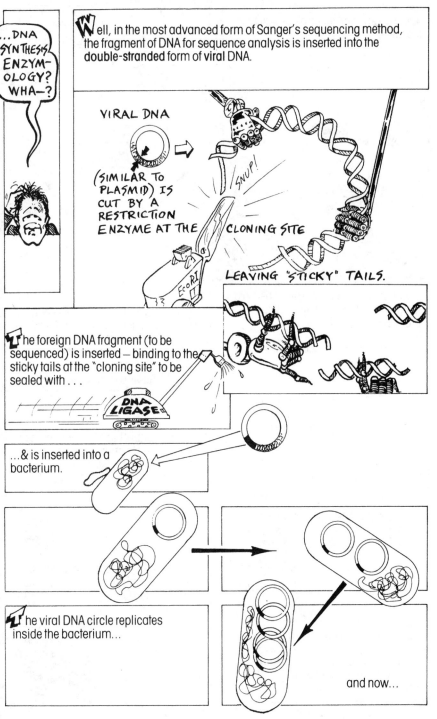

...DNA SYNTHESIS ENZYMOLOGY? WHA—?

Well, in the most advanced form of Sanger's sequencing method, the fragment of DNA for sequence analysis is inserted into the **double-stranded** form of **viral** DNA.

VIRAL DNA

(SIMILAR TO PLASMID) IS CUT BY A RESTRICTION ENZYME AT THE CLONING SITE

SNIP!

EcoRI

LEAVING "STICKY" TAILS.

The foreign DNA fragment (to be sequenced) is inserted — binding to the sticky tails at the "cloning site" to be sealed with . . .

DNA LIGASE

...& is inserted into a bacterium.

The viral DNA circle replicates inside the bacterium...

and now...

125

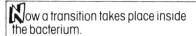 ow a transition takes place inside the bacterium.

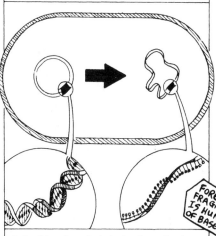

A double-stranded circle gives rise to a single-stranded circle by a complex process!

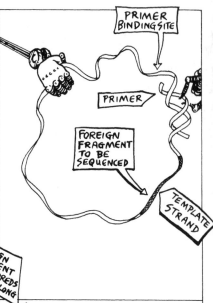

he single-stranded DNA is purified from the bacterium & a short piece of DNA called a **Primer** (15 bases long), whose sequence is exactly complementary to the single strand next to the cloning site, is added.

ere's **DNA polymerase** with its driver. It's specially built to copy single-stranded DNA, but needs a **double-stranded start point**. This is provided by the **Primer**.

ow remember that DNA polymerase is surrounded by many normal free-floating DNA base trucks.

Sanger provided these trucks but also a very few "chain terminating inhibitor trucks" (dideoxy-triphosphates).

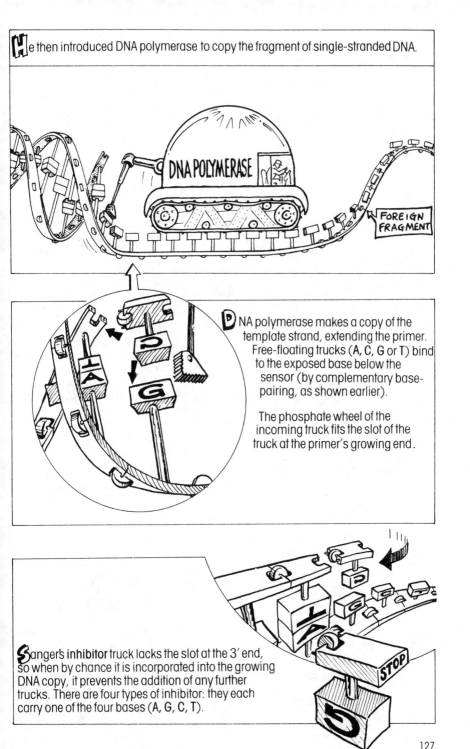

He then introduced DNA polymerase to copy the fragment of single-stranded DNA.

DNA POLYMERASE

FOREIGN FRAGMENT

DNA polymerase makes a copy of the template strand, extending the primer. Free-floating trucks (A, C, G or T) bind to the exposed base below the sensor (by complementary base-pairing, as shown earlier).

The phosphate wheel of the incoming truck fits the slot of the truck at the primer's growing end.

Sanger's **inhibitor** truck lacks the slot at the 3′ end, so when by chance it is incorporated into the growing DNA copy, it prevents the addition of any further trucks. There are four types of inhibitor: they each carry one of the four bases (A, G, C, T).

STOP

127

As the polymerase extends the primer, it adds A, G, T or C trucks as required by base-pairing to the template. Most often when A is required, a normal truck is added & the chain can be extended further. But should an inhibitor be joined, the chain growth stops. DNA polymerase therefore makes pieces of DNA which extend from the primer to each position where A addition is dictated by the base-pairing rules.

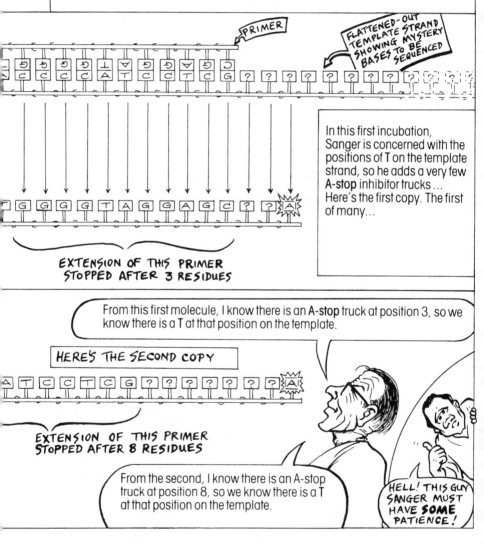

PRIMER

FLATTENED-OUT TEMPLATE STRAND SHOWING MYSTERY BASES TO BE SEQUENCED

In this first incubation, Sanger is concerned with the positions of T on the template strand, so he adds a very few A-stop inhibitor trucks... Here's the first copy. The first of many...

EXTENSION OF THIS PRIMER STOPPED AFTER 3 RESIDUES

From this first molecule, I know there is an A-stop truck at position 3, so we know there is a T at that position on the template.

HERE'S THE SECOND COPY

EXTENSION OF THIS PRIMER STOPPED AFTER 8 RESIDUES

From the second, I know there is an A-stop truck at position 8, so we know there is a T at that position on the template.

HELL! THIS GUY SANGER MUST HAVE **SOME** PATIENCE!

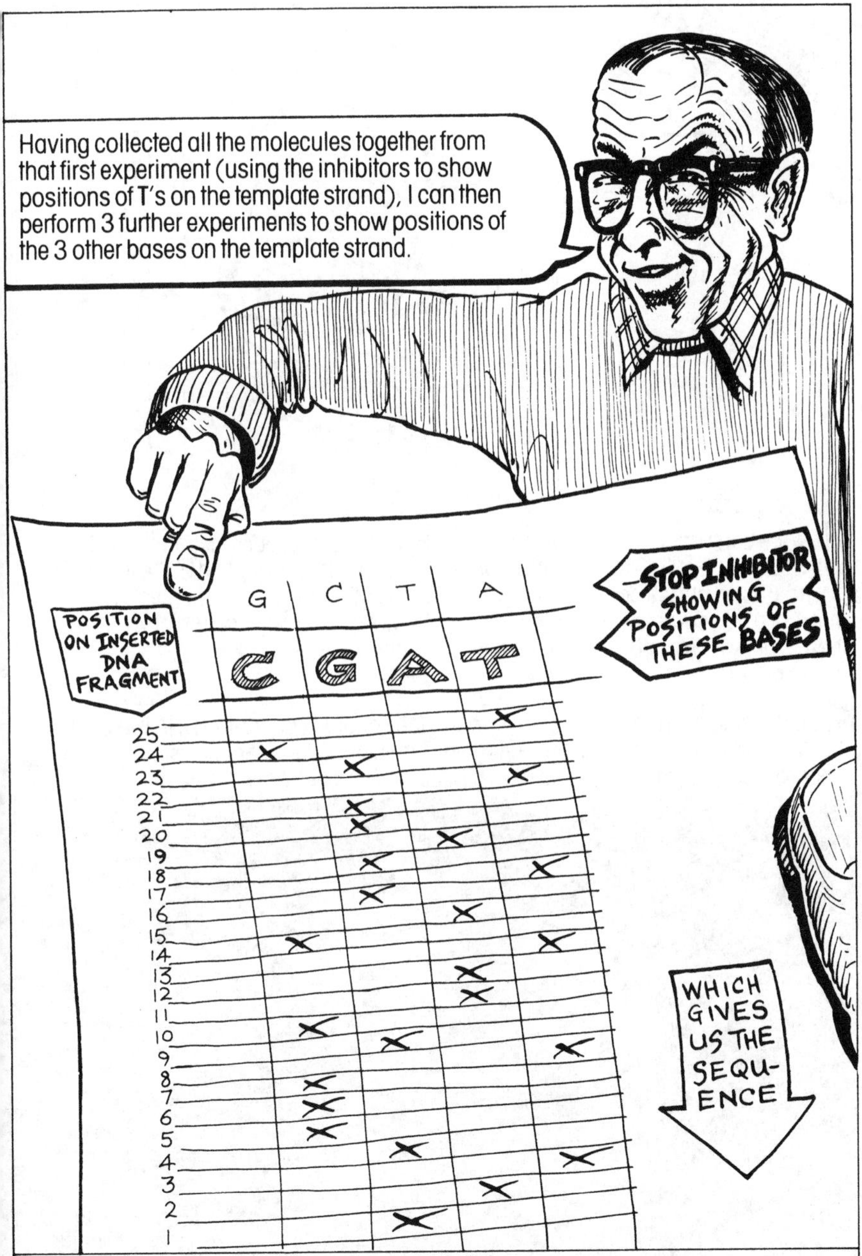
Having collected all the molecules together from that first experiment (using the inhibitors to show positions of T's on the template strand), I can then perform 3 further experiments to show positions of the 3 other bases on the template strand.
POSITION ON INSERTED DNA FRAGMENT
G C T A
C G A T
STOP INHIBITOR SHOWING POSITIONS OF THESE BASES
WHICH GIVES US THE SEQUENCE

POSITION	1	2	3	4	5	6	7	8	9	10	11	12	13	14	15	16	17	18	19	20	21	22	23	24	25
BASE	G	A	T	G	C	C	C	T	G	C	A	A	T	C	A	G	T	G	A	G	G	T	G	C	T

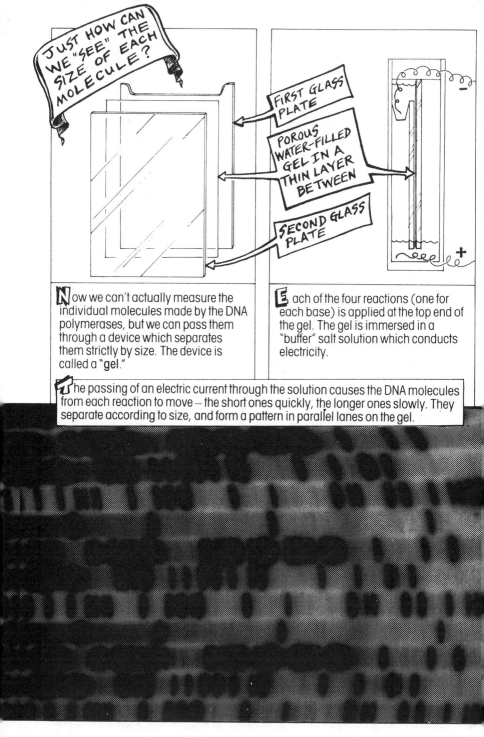

JUST HOW CAN WE "SEE" THE SIZE OF EACH MOLECULE?

FIRST GLASS PLATE

POROUS WATER-FILLED GEL IN A THIN LAYER BETWEEN

SECOND GLASS PLATE

Now we can't actually measure the individual molecules made by the DNA polymerases, but we can pass them through a device which separates them strictly by size. The device is called a "**gel**."

Each of the four reactions (one for each base) is applied at the top end of the gel. The gel is immersed in a "buffer" salt solution which conducts electricity.

The passing of an electric current through the solution causes the DNA molecules from each reaction to move — the short ones quickly, the longer ones slowly. They separate according to size, and form a pattern in parallel lanes on the gel.

...the molecular "rulers" make a pattern on an X-ray film and allow me to read the structure!

The genius of Sanger's method was that the pattern could be interpreted directly to give the complete structure – 1000 bases in a single experiment – of the gene. Walter Gilbert and Alan Maxam of Harvard also devised a method.

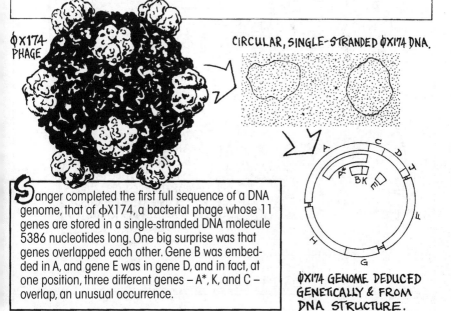

φX174 PHAGE

CIRCULAR, SINGLE-STRANDED φX174 DNA.

Sanger completed the first full sequence of a DNA genome, that of φX174, a bacterial phage whose 11 genes are stored in a single-stranded DNA molecule 5386 nucleotides long. One big surprise was that genes overlapped each other. Gene B was embedded in A, and gene E was in gene D, and in fact, at one position, three different genes – A*, K, and C – overlap, an unusual occurrence.

φX174 GENOME DEDUCED GENETICALLY & FROM DNA STRUCTURE.

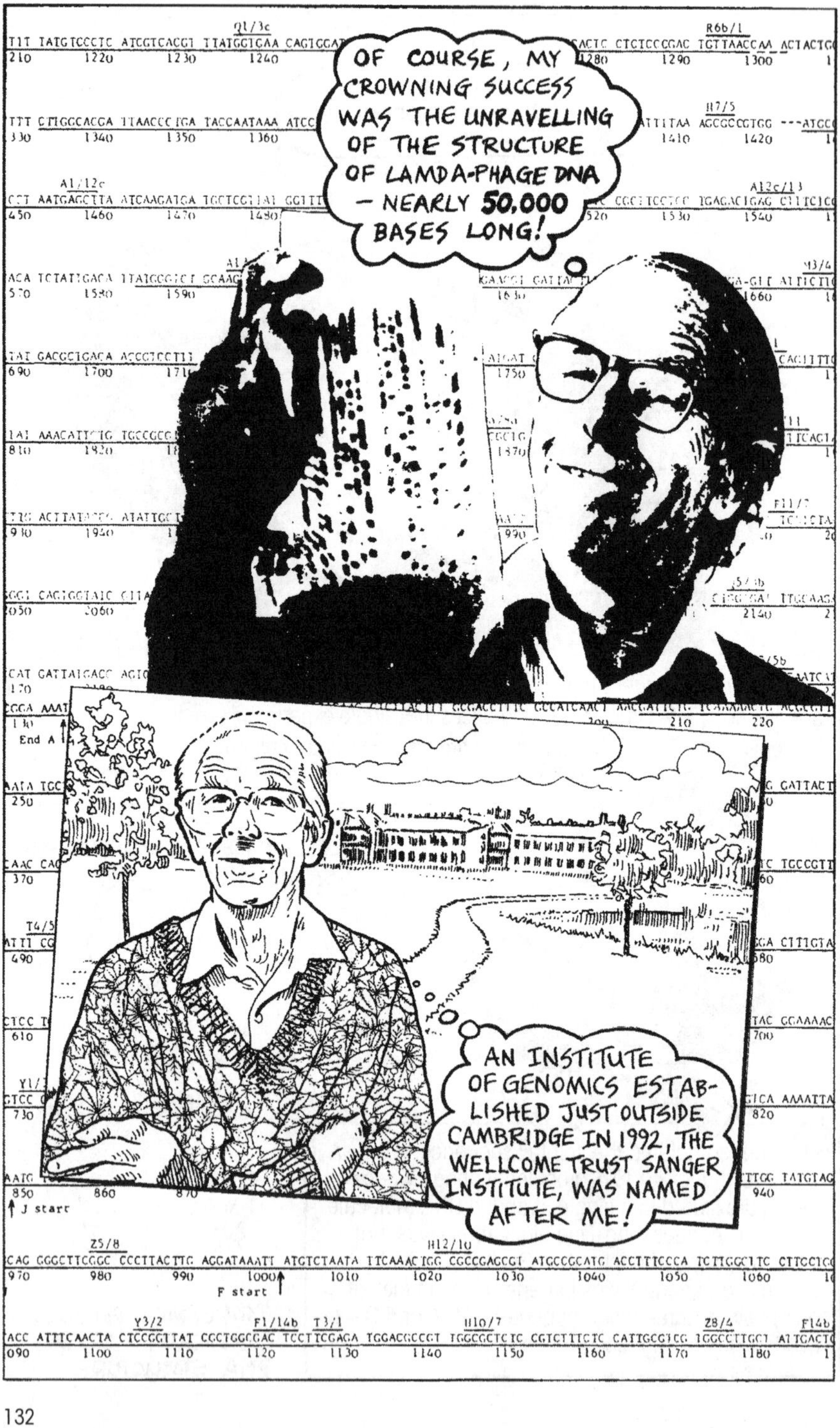
OF COURSE, MY CROWNING SUCCESS WAS THE UNRAVELLING OF THE STRUCTURE OF LAMDA-PHAGE DNA — NEARLY 50,000 BASES LONG!
AN INSTITUTE OF GENOMICS ESTABLISHED JUST OUTSIDE CAMBRIDGE IN 1992, THE WELLCOME TRUST SANGER INSTITUTE, WAS NAMED AFTER ME!

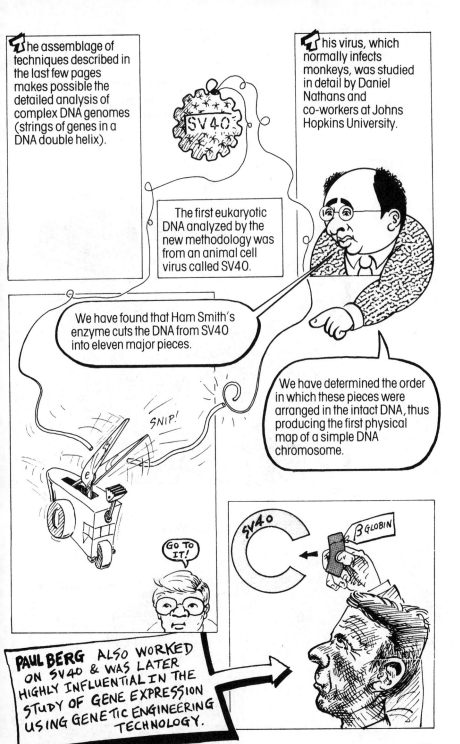

The assemblage of techniques described in the last few pages makes possible the detailed analysis of complex DNA genomes (strings of genes in a DNA double helix).

This virus, which normally infects monkeys, was studied in detail by Daniel Nathans and co-workers at Johns Hopkins University.

The first eukaryotic DNA analyzed by the new methodology was from an animal cell virus called SV40.

We have found that Ham Smith's enzyme cuts the DNA from SV40 into eleven major pieces.

SNIP!

We have determined the order in which these pieces were arranged in the intact DNA, thus producing the first physical map of a simple DNA chromosome.

GO TO IT!

SV40

β GLOBIN

PAUL BERG ALSO WORKED ON SV40 & WAS LATER HIGHLY INFLUENTIAL IN THE STUDY OF GENE EXPRESSION USING GENETIC ENGINEERING TECHNOLOGY.

133

SV40 DNA is a circular molecule 5,200 bases long.

Here are the pieces reassembled.

With the physical map drawn a genetic map of viral DNA could then be developed…

PROTEIN WHICH REPLICATES DNA

Large T antigen

PROTEINS WHICH FORM PROTECTIVE COAT FOR DNA IN VIRUS

ORIGIN OF DNA REPLICATION

Small T antigen

We have now entered the post-Sanger era of genome sequencing (see page 165). First, let's examine the mechanisms by which genes of eukaryotic cells, including those of yeast and animals, are transcribed into RNA and translated into protein. Many of the features of bacterial gene expression that we have already discussed are employed by eukaryotic cells, but the existence of the nucleus required some additional features.

Exons, Introns, and Splicing

Early surveys of eukaryotic DNA revealed that some DNA sequences were present only once per cell. However, others were present many times. Some of the highly repeated sequences were likely to have a **structural** rather than an **informational** role. The coding sequences of genes fell in the "**unique sequence class.**"

For animal cell gene expression, it was messenger RNA and its precursors that were most important. And here the difference with bacteria was dramatic. The first animal cell RNAs studied were from specialized animal cells that synthesize a limited number of proteins, but in great quantity.

Something was known about the RNA of animal cells, too. Like bacteria, animal cells had **messenger RNA** (the train carrying the coded information from RNA polymerase to the protein factory),

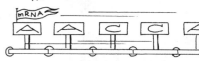

ribosomal RNA (which makes up part of the protein factory),

and **transfer RNA** (which carries amino acids to the protein factory).

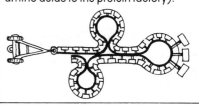

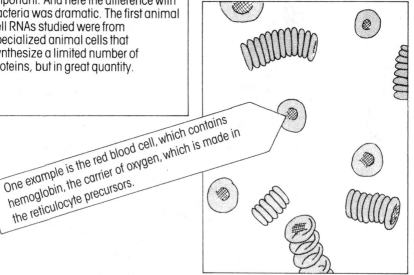

One example is the red blood cell, which contains hemoglobin, the carrier of oxygen, which is made in the reticulocyte precursors.

The protein component, globin, is translated from an mRNA which is plentiful in red blood cells and easily purified.

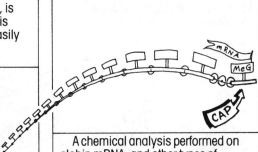

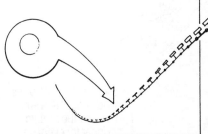

A chemical analysis performed on globin mRNA, and other types of mRNA as well, showed that these molecules had unexpected modifications not found in bacterial mRNA.

At the 5′ end they had an unprecedented "inverted G" residue, called a **cap**. And at the 3′ end they had a long string of A's, up to 200 in number, called "poly A."

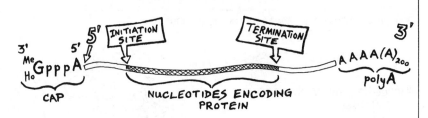

No equivalent to poly A or the cap are in DNA, and these are added to the mRNA by special mechanisms after RNA transcription!

Messenger RNA was made from RNA in the nucleus (nuclear RNA), but how?

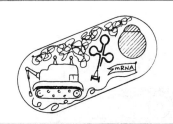

In bacteria, the RNA transcript, as originally copied from DNA, is the mRNA.

But in eukaryotes scientists asked

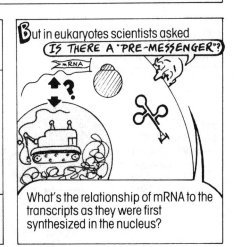

IS THERE A "PRE-MESSENGER"?

What's the relationship of mRNA to the transcripts as they were first synthesized in the nucleus?

Examining the nuclear RNA only added to the confusion.

First there was a lot of it!

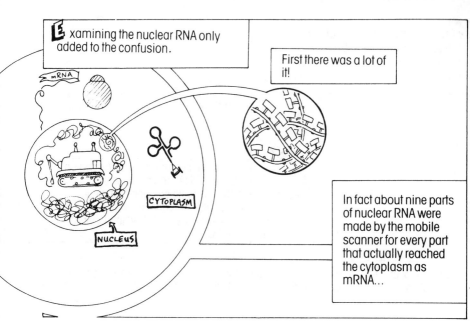

In fact about nine parts of nuclear RNA were made by the mobile scanner for every part that actually reached the cytoplasm as mRNA...

Second, the nuclear RNA was very heterogeneous.

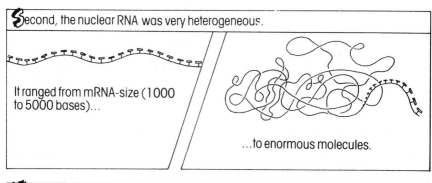

It ranged from mRNA-size (1000 to 5000 bases)...

...to enormous molecules.

How was the first structure of mRNA deduced? By copying it into DNA, then cloning & sequencing:

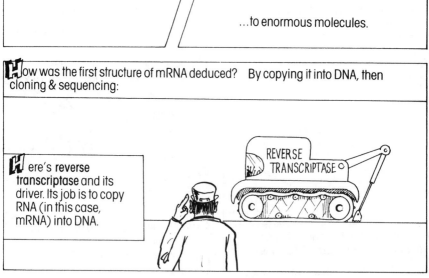

Here's **reverse transcriptase** and its driver. Its job is to copy RNA (in this case, mRNA) into DNA.

REVERSE TRANSCRIPTASE

In the first experiments, reverse transcriptase made copies of mRNA...

Here's how...

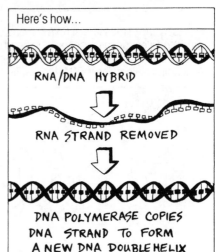

RNA/DNA HYBRID

RNA STRAND REMOVED

DNA POLYMERASE COPIES DNA STRAND TO FORM A NEW DNA DOUBLE HELIX

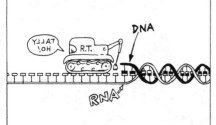

TALLY HO!

R.T.

DNA

RNA

This copy of mRNA was converted to a DNA double helix...

The resulting double-stranded DNA is cloned. By sequencing these clones, the complete structure of globin mRNA was soon known.

This was achieved by a novel application of the hybridization technique in which mRNA was forced to base-pair with its template DNA.

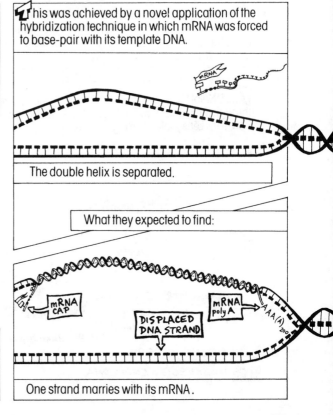

mRNA

The double helix is separated.

What they expected to find:

mRNA CAP

DISPLACED DNA STRAND

mRNA poly A

AAA(A)$_{200}$

The next task was to examine the gene from which the message was copied.

One strand marries with its mRNA.

139

This hybrid of messenger RNA & a template strand of genomic DNA was viewed under an electron microscope … instead of seeing the thick double-stranded hybrid of RNA & DNA – called a **heteroduplex** – with thinner single strands of DNA extending beyond the mRNA, the results were amazing & completely unexpected!

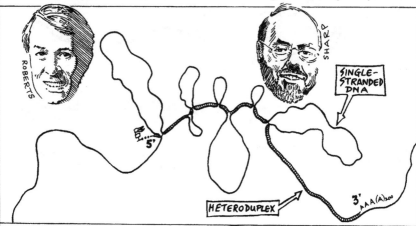

Phil Sharp working at MIT and Richard Roberts working at Cold Spring Harbor saw a much more complex pattern of double and single strands! As explained below, this pattern revealed that to make an mRNA, parts of the primary RNA transcript must be cut out. The DNA is not altered.

The DNA of the genes seemed to be in sections spread through the genome with bits of extraneous DNA in between. Let's look at the β globin gene:

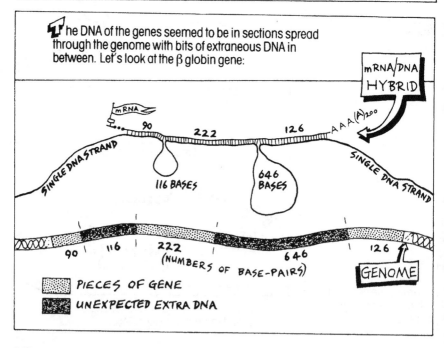

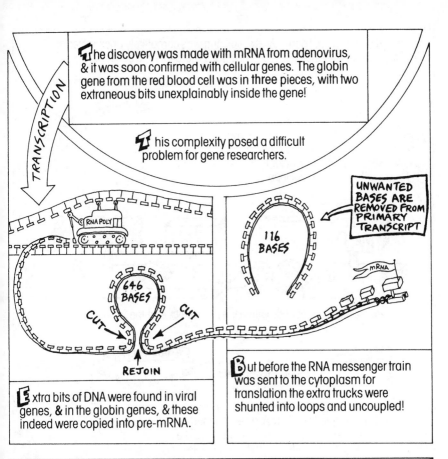

The discovery was made with mRNA from adenovirus, & it was soon confirmed with cellular genes. The globin gene from the red blood cell was in three pieces, with two extraneous bits unexplainably inside the gene!

This complexity posed a difficult problem for gene researchers.

RNA POLY

646 BASES

CUT   CUT

REJOIN

Extra bits of DNA were found in viral genes, & in the globin genes, & these indeed were copied into pre-mRNA.

UNWANTED BASES ARE REMOVED FROM PRIMARY TRANSCRIPT

116 BASES

mRNA

But before the RNA messenger train was sent to the cytoplasm for translation the extra trucks were shunted into loops and uncoupled!

TRANSCRIPTION

We can imagine the primary transcript as a manuscript with sections of gibberish in the text. Garbled parts have to be removed by an editor before the book can be published.

The reaction which ~~grn splk hog uni spoilg~~ch removed the extra sequences (called **introns**) and re~~gazo oglytr ynpcehle~~joined the mRNA seque~~prenurgle astand tweeple~~ nces (called exons) ~~aras nismaint~~ is called **RNA Splicing**.

The reaction which removed the extra sequences (called introns) and rejoined the mRNA sequences (called exons) is called **RNA Splicing**.

The existence of split genes & **RNA splicing** was completely unexpected! It reaffirmed the difference between eukaryotic & prokaryotic genes...

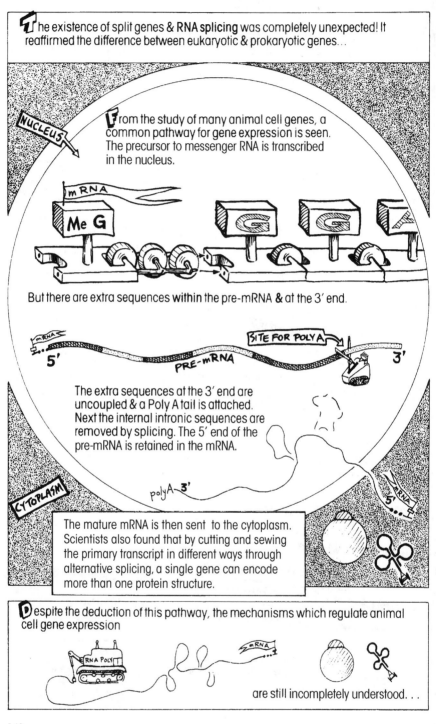

From the study of many animal cell genes, a common pathway for gene expression is seen. The precursor to messenger RNA is transcribed in the nucleus.

NUCLEUS

mRNA

Me G   G   G

But there are extra sequences **within** the pre-mRNA & at the 3' end.

mRNA
5'   PRE-mRNA   SITE FOR POLY A   3'

The extra sequences at the 3' end are uncoupled & a Poly A tail is attached. Next the internal intronic sequences are removed by splicing. The 5' end of the pre-mRNA is retained in the mRNA.

CYTOPLASM

polyA-3'

mRNA
5'...

The mature mRNA is then sent to the cytoplasm. Scientists also found that by cutting and sewing the primary transcript in different ways through alternative splicing, a single gene can encode more than one protein structure.

Despite the deduction of this pathway, the mechanisms which regulate animal cell gene expression

RNA POLY   mRNA

are still incompletely understood. . .

142

CHROMATIN AND HISTONES

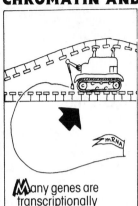

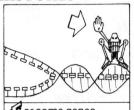

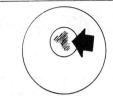

Many genes are transcriptionally controlled.

For some genes repressors or activators which bind to the promoter are likely to control transcription, just as in bacterial genes.

However for many eukaryotic genes, the structure of the **chromatin** (dark-staining material inside the nucleus) may be critical and controlled by other proteins.

Chromatin is a complex of eukaryotic DNA with positively charged proteins called **histones**.

DNA winds twice about the core to form the fundamental subunit of chromatin called the **nucleosome.**

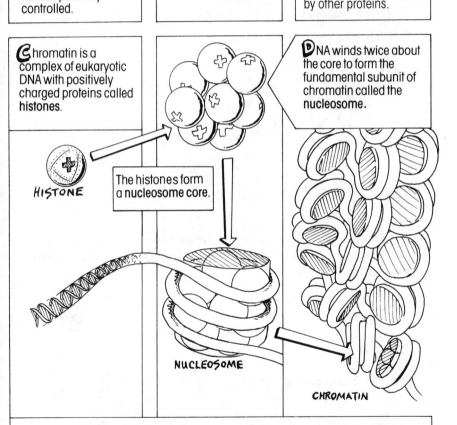

HISTONE

The histones form a **nucleosome core.**

NUCLEOSOME

CHROMATIN

Chromatin consists of many nucleosomes linked by DNA & packaged into more complex but regular fibers. The structure of the chromatin differs, depending on the activity of the gene. Open conformations of chromatin are ready for transcription, while inactive chromatin is closed to the surroundings & compact.

GENE FAMILIES

Different genes are expressed as cells differentiate.

During development, different globin proteins are expressed: the embryonic, the fetal, and finally the adult globin. The developing organism's requirements for transporting oxygen change as it grows from embryo to fetus to adult. Therefore different forms of the oxygen-carrying globin protein are produced through the successive activation of genes for:

1. **Embryonic** (up to 12 weeks)   2. **Fetal** (up to birth)   3. **Adult** globin (from birth)

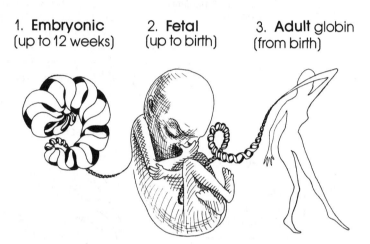

The genes for the different globin types are situated together within a 40,000-base region of the human genome. They form a **gene family**.

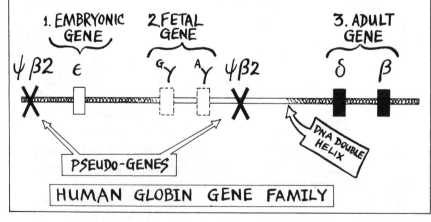

HUMAN GLOBIN GENE FAMILY

Now let's put the human globin genes into context with similar genes in other primates.

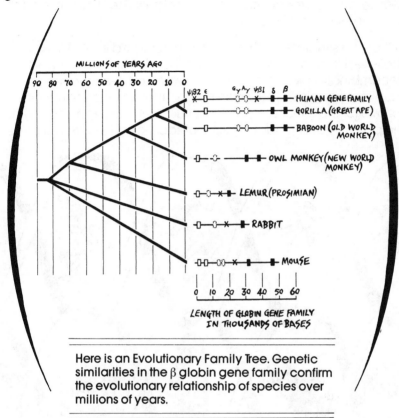

MILLIONS OF YEARS AGO

Here is an Evolutionary Family Tree. Genetic similarities in the β globin gene family confirm the evolutionary relationship of species over millions of years.

Although embryonic, fetal, and adult genes are different and suited to the specific stages of development, nucleotide sequences show them to be closely related. Perhaps they had a primordial gene as a common ancestor. Duplication of this primordial gene would have allowed the separate copies to evolve to their current structures.

Changes in chromatin structure of this globin cluster during development may switch expression from embryonic to fetal to adult globin. The globin cluster also contains a non-fuctional relic of a globin gene which was inactivated through mutation. This is called a globin pseudo-gene.

Controlling Genes for Antibodies

Some genes are controlled by unusual mechanisms, such as genes for antibodies.

When the structure of DNA was first elucidated in 1953, it was believed that random mutations in the DNA structure and sexual recombination would account for evolution. Genes often exist in duplicate copies in an organism and the process of duplication allows for the creation of mutant structures – that may or may not help the organism adapt – without sacrificing the original gene.

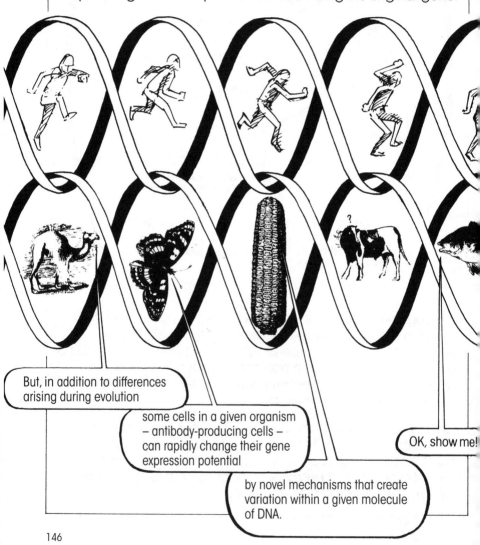

But, in addition to differences arising during evolution

some cells in a given organism – antibody-producing cells – can rapidly change their gene expression potential

by novel mechanisms that create variation within a given molecule of DNA.

OK, show me!

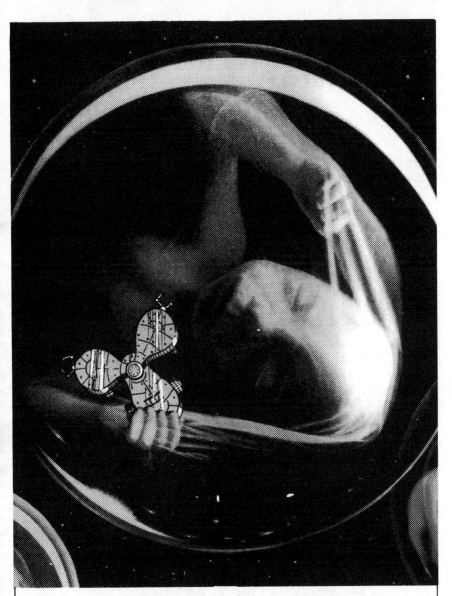

It was soon learned that there are mechanisms for rearranging DNA within a particular plant or animal. One of the most remarkable (and unusual) of these is responsible for the production of millions (at least) of **antibodies** from a few hundred antibody genes, permitting man to survive when infected by new kinds of organisms. The stock of possible antibodies is determined throughout life.

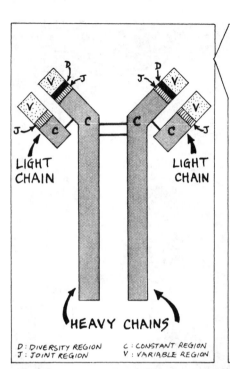

Antibodies consist of two types of protein chain: light and heavy.

Each light chain protein has variable and constant regions. The variable consists of two segments: V and J. The constant is called C. Heavy chains are similar except their variable region is in three segments: V, D and J; as well as a constant C region.

LIGHT CHAIN   LIGHT CHAIN

HEAVY CHAINS

D : DIVERSITY REGION   C : CONSTANT REGION
J : JOINT REGION   V : VARIABLE REGION

The remarkable property of the immune system is its ability to create an immense number of antibody specificities. The variety results from the differences in the variable regions in the light and heavy chains.

How does this arise?

In the embryonic DNA we find the genes for these various segments. But there are many of each kind of segment (D, V, or J) and they are widely separated in the DNA, that is, **not yet assembled as a functional gene.** To make the gene for a light chain or a heavy chain, the distant DNA segments must be brought together and joined.

There are many possibilities for this joining:

ABOUT 40-70
V GENES

ABOUT 25
D GENES

ABOUT 6
J GENES

ONE **C** GENE
FOR FIRST
JOINING STEP

Here's a heavy chain family.

In precursors to antibody-producing cells, an active heavy chain gene is made by joining one **D** gene to make a **DJ** (about 20 x 6 = about 150 possibilities), and then the **DJ** to a **V** gene to made a **VDJ** gene (about 50 x 150 = 7500 possibilities).

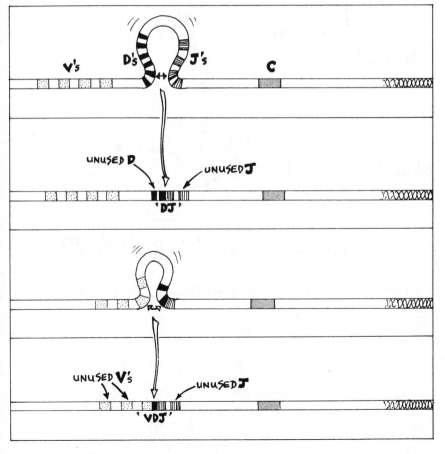

149

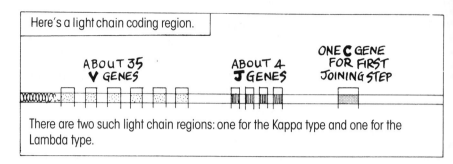

Here's a light chain coding region.

ABOUT 35
V GENES

ABOUT 4
J GENES

ONE C GENE
FOR FIRST
JOINING STEP

There are two such light chain regions: one for the Kappa type and one for the Lambda type.

For light chains a similar DNA rearrangement brings **V** to **J**. The possibilities for different light chains are about 35 x 4 = 140 for Kappa, and about the same for Lambda.

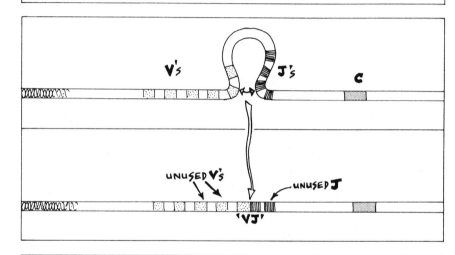

V's   J's   C

UNUSED V's   UNUSED J

'VJ'

Even after such DNA rearrangement, the **VDJ** for heavy chains is separated from the heavy chain **C** gene; also the **VJ** for light chains is separated from the light chain **C** gene.

For both heavy and light, **C** is brought into place **after** transcription by the **splicing** step, which forms the mature mRNA.

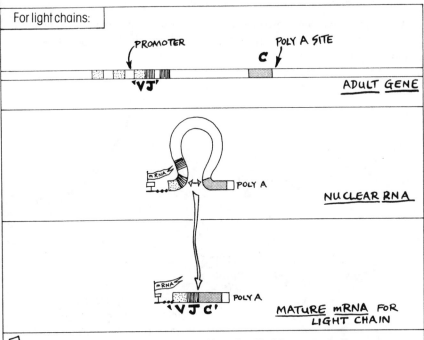

For light chains:

PROMOTER

POLY A SITE

C

'VJ'

ADULT GENE

mRNA

POLY A

NUCLEAR RNA

mRNA

POLY A

'VJC'

MATURE mRNA FOR LIGHT CHAIN

For heavy chains the **VDJ** RNA transcript is spliced to C in a very similar manner so that **VDJC** sequences are adjacent in final mRNA.

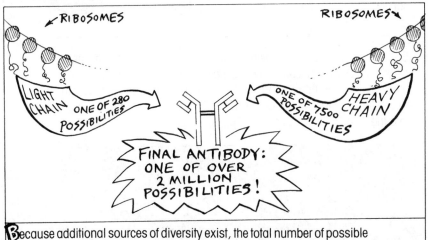

RIBOSOMES

RIBOSOMES

LIGHT CHAIN ONE OF 280 POSSIBILITIES

ONE OF 7500 POSSIBILITIES

HEAVY CHAIN

FINAL ANTIBODY: ONE OF OVER 2 MILLION POSSIBILITIES!

Because additional sources of diversity exist, the total number of possible antibodies is, in fact, higher. The most important is mutation of variable regions during cell division (somatic hypermutation).

Mature mRNA's for light and heavy chains are thus translated.

After translation in the ribosome, the light & heavy protein chains form an antibody ...

... which is sent to the surface of the cell which produced it.

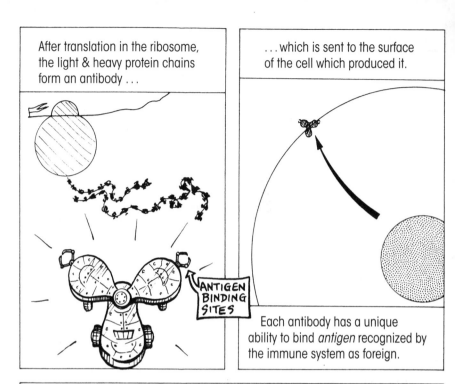

ANTIGEN BINDING SITES

Each antibody has a unique ability to bind *antigen* recognized by the immune system as foreign.

Perched on the cell surface the antibody surveys for any antigens it can bind.

Each precursor to an antibody-producing cell has recombined its antibody genes in a slightly different way, making different antibodies which bid different antigens. Remarkably, binding of antigen triggers the producing cell to proliferate to tens of thousands of identical cells, each spewing antibody out into the bloodstream!

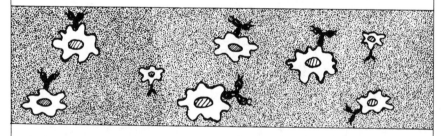

In this way rearranging antibody genes creates diversity, and antigen binding selects the necessary response.

Other mechanisms depending on DNA rearrangement may modify the structure of the genome. Transposable elements (jumping genes) have been found that can insert themselves into a variety of sites in DNA, causing mutations, inversions, and the turning on or off of genes.

Even the so-called "junk" – the DNA in the introns that does not appear to have any clear function – has been found to play an important role in regulation and the production of variation.

NEWLY PLACED GENE

As we will see, in the 1940s, Barbara McClintock discovered "jumping genes" (see page 223). Jumping genes are capable of moving from place to place within the chromosome and inserting themselves between or within other genes. In some cases genes which neighbor the insertion site may be turned on or off when a new gene jumps into their vicinity. When the jumping gene is inserted within another gene, it can alter the coded information of that gene. Jumping genes were first discovered by McClintock in maize, and it is known that certain genes in animal cells, such as the immunoglobulin genes in mammals, also "jump" through highly controlled DNA rearrangements. Some DNA sequences jump by making copies of themselves which can insert at other sites in the chromosome. These DNAs "proliferate" within the chromosome through the jumping mechanism, and may be found repeated hundreds of thousands of times within the genome.

CHROMOSOMES

Each chromosome is a single double-stranded DNA, which in humans is a linear molecule. If fully stretched, the DNA of a typical chromosome would be about 2 yards in length.

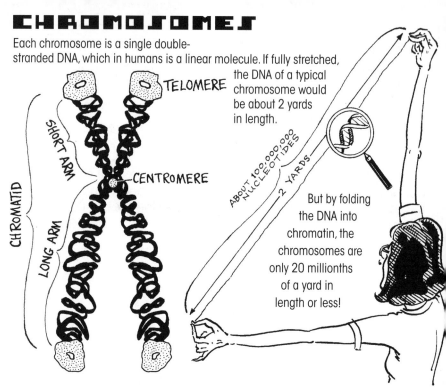

TELOMERE

CENTROMERE

SHORT ARM

LONG ARM

CHROMATID

ABOUT 100,000,000 NUCLEOTIDES

2 YARDS

But by folding the DNA into chromatin, the chromosomes are only 20 millionths of a yard in length or less!

Chromosomes have arms (chromatids) that are joined at the center by the centromere.

While goldfish have 47 kinds of chromosomes per cell and dogs have 39, humans have 23 kinds, each present in two copies, except for X and Y in males, of which there are one each. The Y chromosome, found only in males, is the smallest chromosome with about 58,000,000 nucleotides, while chromosome #1 is the largest with over 247,000,000 nucleotides.

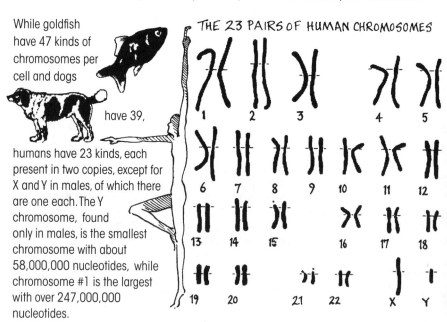

THE 23 PAIRS OF HUMAN CHROMOSOMES

1 2 3 4 5
6 7 8 9 10 11 12
13 14 15 16 17 18
19 20 21 22 X Y

TELOMERES

The DNA in chromosomes is one long, linear, double-stranded molecule. DNA polymerase replicates the majority of the DNA; however, it cannot complete the job. It has no trouble synthesizing one strand of the double helix (the "leading strand"), which is made continuously until the polymerase reaches the end of the strand. But the polymerase cannot finish the other strand, which is called the lagging strand. The lagging strand is made in many short segments, each started by an RNA "primer" (see page 58). When the last RNA primer is removed, the place where it was bound remains uncopied and thus the lagging strand remains incomplete. This problem arises within the telomeres, structures at the ends of chromosomes about 100,000 bases in length. Telomeres consist of short repeat DNA sequences – in vertebrates a TTAGGG sequence repeated over and over again. Elizabeth Blackburn and Carol Greider discovered a specialized DNA polymerase called telomerase that replicates telomeres by copying an RNA template. Telomerase provides the critical step in maintaining chromosome length. Without the telomerase mechanism, the chromosomes would shorten each time the DNA is replicated. If shortening takes place, the cells age. However, if telomerase is overactive, cancer can result.

Scientists hypothesize that if the telomerase could be reactivated, the cellular aging process could be halted.

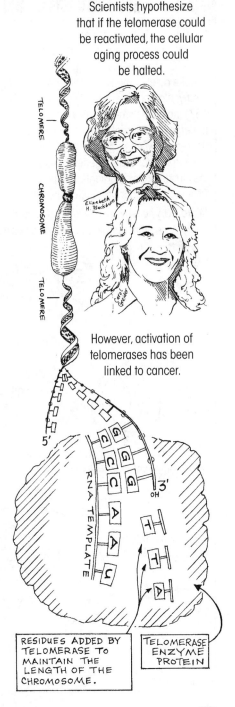

However, activation of telomerases has been linked to cancer.

RESIDUES ADDED BY TELOMERASE TO MAINTAIN THE LENGTH OF THE CHROMOSOME.

TELOMERASE ENZYME PROTEIN

Imprinting and Micro RNAs: "Hidden" Layers of Gene Regulation

Imprinting. Some genes are expressed differently if inherited from mother or father. Chromosomes of eggs or sperm acquire a set of marks (either DNA methylation or histone acetylation; see page 157) indicating whether they are of maternal or paternal origin. These marks are erased when eggs and sperm are created in subsequent generations. After they fuse, as development proceeds, new DNA methylation marks are created.

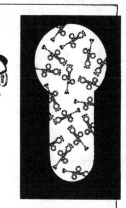

Micro RNA activity. In 1998 Andy Fire and Craig Mello studied how small RNAs block gene expression in the nematode worm.

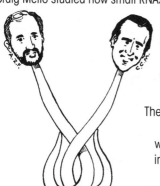

"To our surprise we found that double-stranded RNA was substantially more effective at producing interference than was either strand individually."

They had uncovered an unconventional mechanism of gene control by "RNA interference" (RNAi), in which small molecules of RNA about 20 nucleotides in length, called micro RNA, regulate gene activity by destroying mRNA. Micro RNAs may provide a hitherto hidden RNAi code for gene silencing that fine-tunes gene expression.

These discoveries also revealed the role of small RNAs in disease and the possibility of using some synthetic RNAs, siRNAs, to silence unwanted genes.

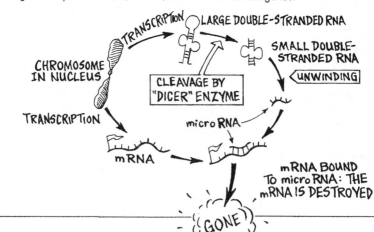

Epigenetics

Epigenetics is the transmission, from one cell to its descendants, of genetic information not encoded in the sequences of nucleotides of the DNA.

Epigenetic mechanisms include DNA methylation, imprinting and micro RNA activity. **DNA methylation:** a carbon atom with three hydrogens (CH_3, a methyl group) is added to one of the bases, usually a cytosine that lies next to a guanine in the DNA strand, a sequence written "CpG":

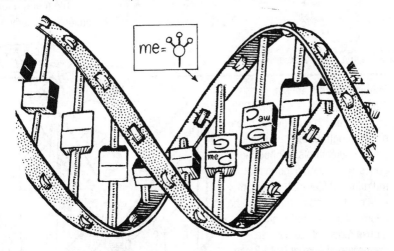

me =

During replication of a DNA molecule with methyl C (MeC), a complementary CpG lacking the methyl groups is synthesized in the daughter strand. An enzyme, DNA methylase, adds the methyl group to the new C residues of the daughter strand CpG. In this way the methyl group is "inherited."

MeCpG is a sequence that frequently appears in clusters called CpG islands. Proteins that bind to MeCpG may silence a neighboring gene. Silencing depends on placing the DNA into a compact chromatin structure. The more methyl groups attached to a region of the DNA, the more likely the gene will be silenced. This is an epigenetic effect because DNA sequences are not changed by methylation. The methylation mechanism:

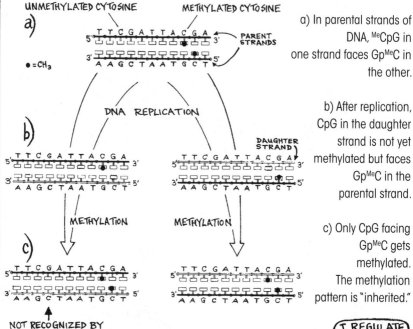

a) In parental strands of DNA, MeCpG in one strand faces GpMeC in the other.

b) After replication, CpG in the daughter strand is not yet methylated but faces GpMeC in the parental strand.

c) Only CpG facing GpMeC gets methylated. The methylation pattern is "inherited."

Epigenetics and Autism.
A gene regulatory protein, MeCP2, emphasizes the importance of epigenetic control. MeCP2 binds to MeC's and controls neighboring genes. Huda Zoghbi found that mutations of MeCP2 that disrupt DNA binding cause Rett Syndrome, a form of autism. Strikingly, in a mouse model of the disease, expression of normal MeCP2 can reverse the symptoms, giving hope for Rett treatment.

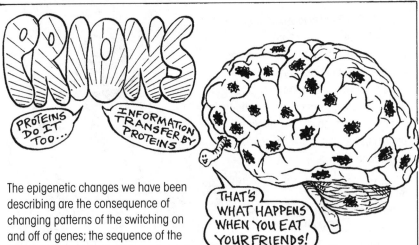

PRIONS

PROTEINS DO IT TOO...

INFORMATION TRANSFER BY PROTEINS

THAT'S WHAT HAPPENS WHEN YOU EAT YOUR FRIENDS!

The epigenetic changes we have been describing are the consequence of changing patterns of the switching on and off of genes; the sequence of the DNA is not altered even if the epigenetic changes can be passed from one generation to the next. There are also non-genetic forms of inheritance. Best known is the prion, a protein structure that does not contain DNA or RNA, and that is associated with a number of nervous disorders in cows (bovine spongiform encephalopathy – "mad cow" disease), sheep (scrapie), and humans (Creutzfeldt-Jakob disease).

The disease, first noticed among the Fore people of New Guinea (called Kuru in New Guinea), causes shaking, trembling, blurred speech and other behavioral disabilities, eventually resulting in death. Women and children were most affected by the disease. The American virologist Carlton Gajdusek demonstrated that injection of brain tissue from those who had died of the disease into the brains of chimpanzees caused the disease in the chimps.

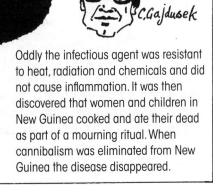

C.Gajdusek

Oddly the infectious agent was resistant to heat, radiation and chemicals and did not cause inflammation. It was then discovered that women and children in New Guinea cooked and ate their dead as part of a mourning ritual. When cannibalism was eliminated from New Guinea the disease disappeared.

In the 1980s Stanley Prusiner proposed that a protein that underwent an abnormal change of shape causes the disease. The abnormal form of the protein causes other normal molecules of this protein to change to the abnormal form. More and more of the abnormal protein accumulates and causes damage to brain cells.

Since the transmission of the abnormal shape of the protein does not involve DNA, this is an epigenetic phenomenon.

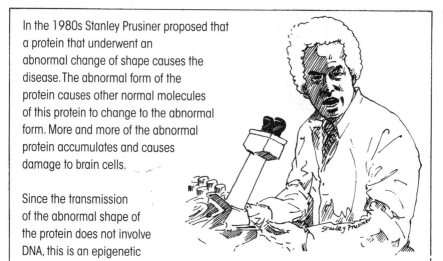

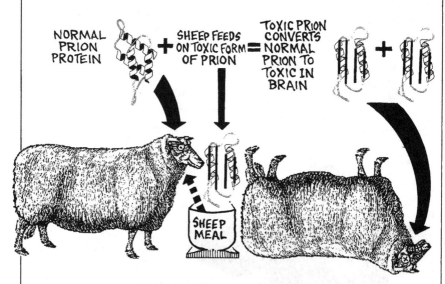

NORMAL PRION PROTEIN + SHEEP FEEDS ON TOXIC FORM OF PRION = TOXIC PRION CONVERTS NORMAL PRION TO TOXIC IN BRAIN

SHEEP MEAL

In England in the 1980s, mad cow disease broke out. Cattle had been fed sheep and cattle protein supplements (cattle and sheep cannibalism) that contained the abnormal proteins. The infected cattle were fed to other cattle and the disease spread. Humans also ate the cattle and some contracted the disease. Protein misfolding, such as in the case of prions, may cause other neurological diseases, including Alzheimer's disease.

THE
HUMAN GENOME PROJECT

On June 26, 2000, President Clinton in Washington and Prime Minister Blair in London simultaneously announced the first draft of the Human Genome.

PRESENTING....

THE HUMAN GENOME!

THE HUMAN GENOME

The entire human genome was sequenced by the International Human Genome Sequencing Consortium: the NIH Human Genome Project headed by Francis Collins. It cost about $3 billion in public funds. Celera Genomics, headed by Craig Venter, sequenced the genome for about $300 million but Celera made extensive use of human genome structure that was in the public domain. The Human Genome Consortium reported its sequence on February 15, 2001, in the journal *Nature* and the Celera Genomics team's sequence appeared the following day in *Science* magazine. Each paper had scores of authors.

WEAK CELERY PUN

Patenting genes: Many scientists decoded parts of the human genome and some tried to patent the sequences they found. Applications have been made for millions of DNA patents. A company, Human Genome Sciences, founded by William Haseltine, a noted Harvard professor, tried to patent hundreds of thousands of gene fragments. This infuriated many scientists, who said that only complete genes that have been isolated and modified to a form not found in nature should be patentable.

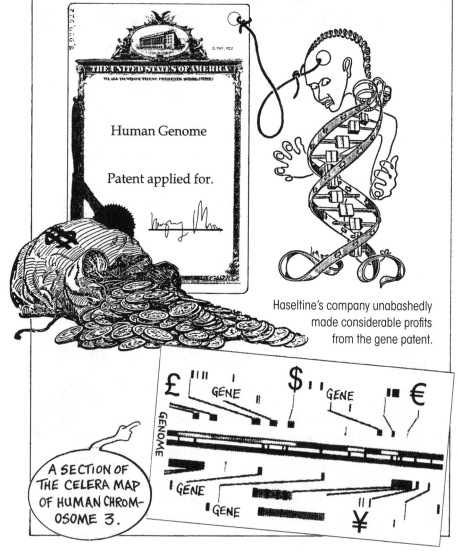

Haseltine's company unabashedly made considerable profits from the gene patent.

A SECTION OF THE CELERA MAP OF HUMAN CHROMOSOME 3.

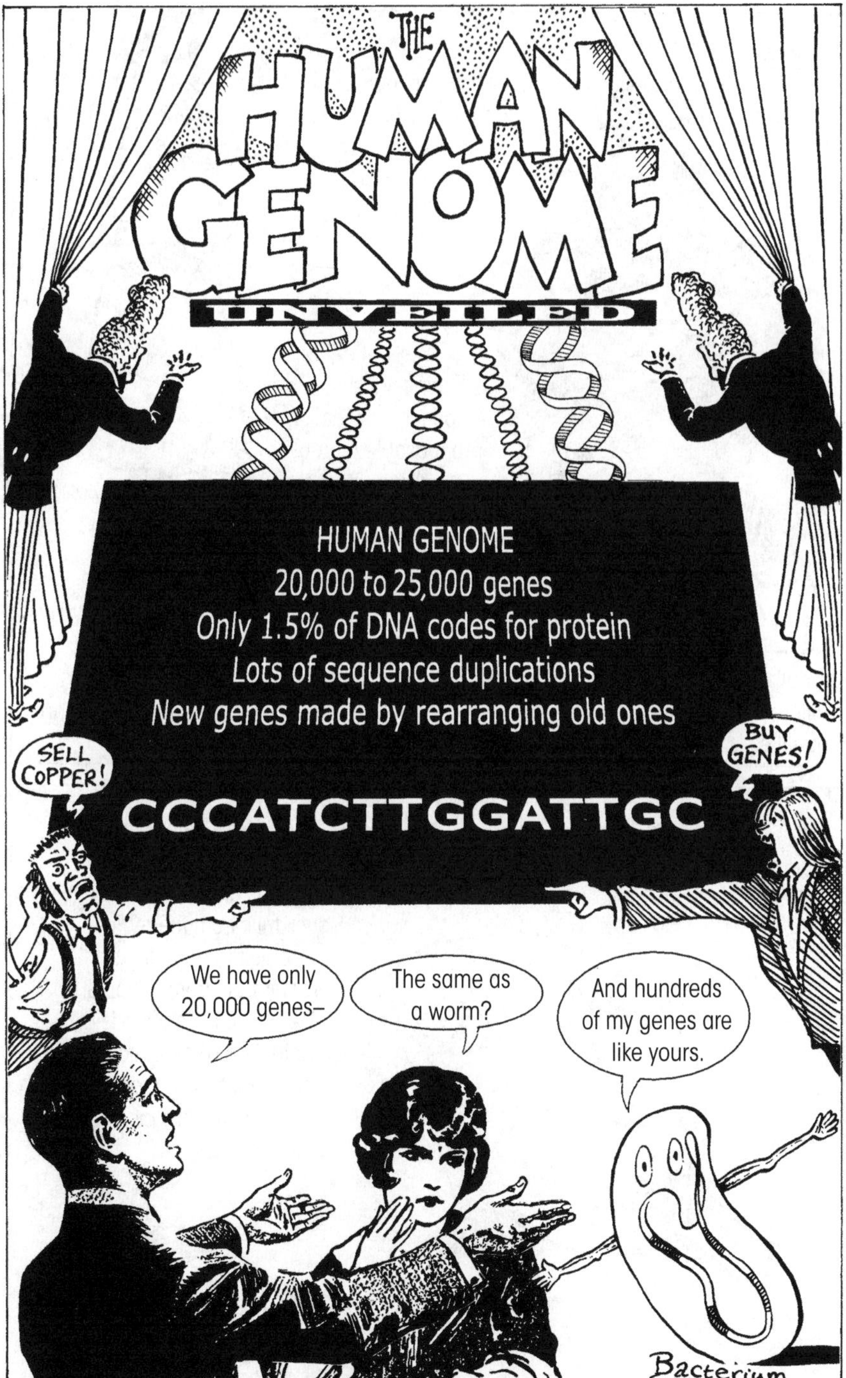
THE HUMAN GENOME
UNVEILED
HUMAN GENOME
20,000 to 25,000 genes
Only 1.5% of DNA codes for protein
Lots of sequence duplications
New genes made by rearranging old ones
SELL COPPER!
BUY GENES!
CCCATCTTGGATTGC
We have only 20,000 genes–
The same as a worm?
And hundreds of my genes are like yours.
Bacterium

I have as many genes as you, and our genomes have greater than 95% similarity.

Yes, but my genes give me:
-speech and reason
-opposable thumbs
-walking upright
-a larger brain, and more.

Small differences between the genomes of humans and chimpanzees make a big difference in physical and mental characteristics. But the basis for the special human qualities is not known.

The platypus, which is midway between reptile and mammal, has a novel genome. The platypus lays eggs and the newborn suck milk from the mother, albeit through the skin rather than from a nipple. The platypus genome is a combination of mammalian and reptilian sequences, an oddity in evolution.

HIGH THROUGHPUT SEQUENCING

Biology entered the era of "high throughput" sequencing, and new "massively parallel" methods can now determine millions of nucleotides in a single analysis. This gave rise to the new sciences of genomics and proteomics, in which computers extract massive quantities of data about DNA and proteins from genome sequences. One goal is to sequence an entire human genome for $1000. Personalized whole genome sequencing will guide the treatments for heart disease and neurodegenerative diseases. New cancer drugs will treat specific cancer-prone genetic conditions, based on genome sequences, and soon, full genomic analysis may be an essential step in treatment of many diseases.

HIGH THROUGHPUT SEQUENCER

THERE YOU ARE, MR SMITHERS, YOUR DNA SEQUENCES & DIAGNOSIS.

THAT'LL DO NICELY.

M.D.

But how can we map genes for genetic diseases?

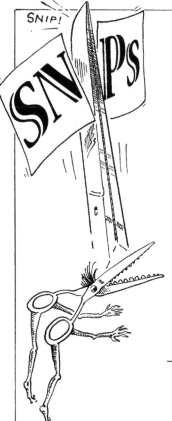

Scientists have developed novel gene-mapping tools that speed the process of finding the gene responsible for a genetic disease, or for comparing two individuals genetically or establishing an individual's ancestral origins. One example involves the study of single nucleotide polymorphisms.

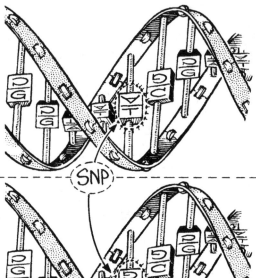

Single nucleotide polymorphisms (SNPs, pronounced "snips") are changes in single DNA residues of the genome that are fairly common. A change is considered a SNP if it is found in at least 1% of the population. Over two million SNPs have been identified and they are found on all chromosomes, throughout the human genome. SNPs are powerful tools for assessing if an individual is at risk for a genetic ailment. An example is the apolipoprotein E gene, *ApoE*, which plays a role in inherited forms of Alzheimer's disease. Using SNPs, we can tell if a person has inherited a form of the *ApoE* gene that increases risk of the disease. SNPs may also reveal a genetic disposition to other diseases, such as cancer or schizophrenia, or even a predisposition to substance abuse.

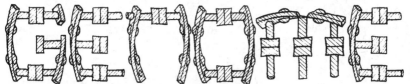

Manipulating the GENOME

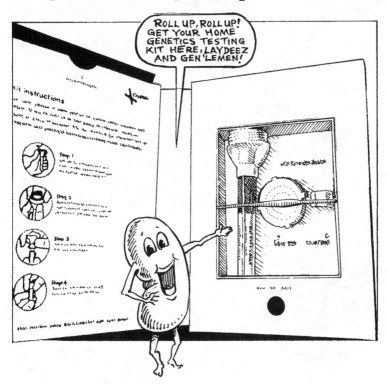

Genetic medicines for genetic diseases

Gene mutation can cause hereditary disease. Of the 20,000 to 25,000 human genes, mutations in about 1,800 genes have been linked to specific diseases. These diseases, many of which result from the mutation within a single gene, may eventually be treated using new "genetic medicines," such as stem cells and novel DNA and RNA drugs consisting of short nucleic acid fragments. These novel drugs can control the activity or expression of genes by modifying the RNAs they encode, most often by modifying mRNAs. Genetic medicines also include DNA or RNA that is introduced into tissues to alter gene expression (see siRNA, page 156).

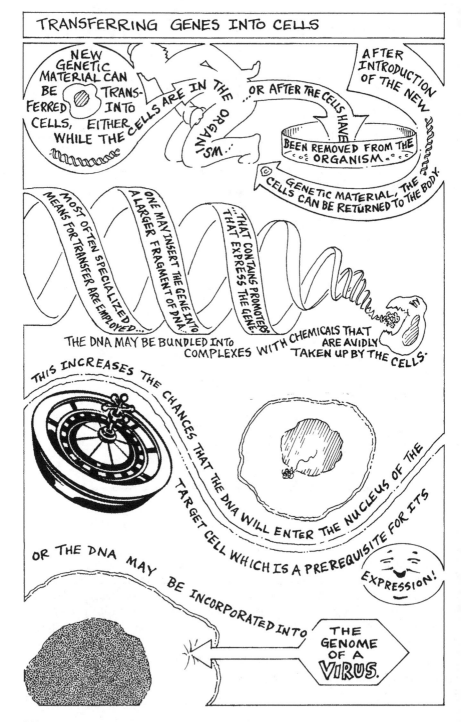

NEW GENETIC MATERIAL CAN BE TRANSFERRED INTO CELLS, EITHER WHILE THE CELLS ARE IN THE ORGANISM... ...OR AFTER THE CELLS HAVE BEEN REMOVED FROM THE ORGANISM.

AFTER INTRODUCTION OF THE NEW GENETIC MATERIAL, THE CELLS CAN BE RETURNED TO THE BODY.

MOST OFTEN SPECIALIZED MEANS FOR TRANSFER ARE EMPLOYED... ONE MAY INSERT THE GENE INTO A LARGER FRAGMENT OF DNA... ...THAT CONTAINS PROMOTERS THAT EXPRESS THE GENE.

THE DNA MAY BE BUNDLED INTO COMPLEXES WITH CHEMICALS THAT ARE AVIDLY TAKEN UP BY THE CELLS.

THIS INCREASES THE CHANCES THAT THE DNA WILL ENTER THE NUCLEUS OF THE TARGET CELL WHICH IS A PREREQUISITE FOR ITS EXPRESSION!

OR THE DNA MAY BE INCORPORATED INTO THE GENOME OF A VIRUS.

TRANSFERRING GENES INTO ANIMALS

Since ancient times, animals have been genetically modified through breeding. Producing the desired characteristics required generations of breeding, but now it is possible to make the changes of one's choice in the genome in a single step.

There are several types of genetic manipulations of animals. Transgenic mice are created when a foreign gene is added to the mouse's genome, resulting in the expression of a new protein.

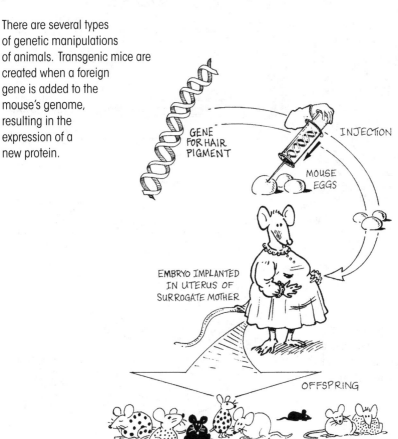

Transgenic mice

To make a transgenic mouse, a copy of the DNA containing the gene one desires to express is injected into fertilized mouse eggs, which are reimplanted into the uterus of a surrogate mother. Some of the offspring incorporate the foreign DNA into their genomes and express the desired protein.

REPLACING GENES IN ANIMALS

In a second type of genetic manipulation, pioneered by Mario Capecchi, a specific gene is inactivated, often by eliminating its DNA altogether from the genome. The result is called a knock-out mouse. It works this way:

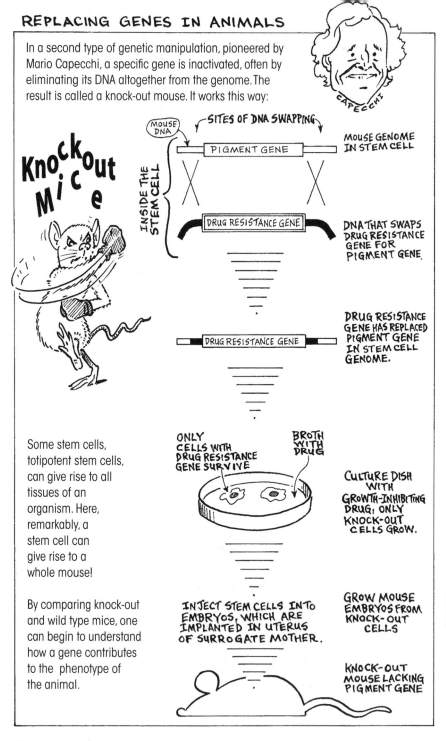

Knockout Mice

MOUSE DNA — SITES OF DNA SWAPPING

INSIDE THE STEM CELL

PIGMENT GENE — MOUSE GENOME IN STEM CELL

DRUG RESISTANCE GENE — DNA THAT SWAPS DRUG RESISTANCE GENE FOR PIGMENT GENE.

DRUG RESISTANCE GENE — DRUG RESISTANCE GENE HAS REPLACED PIGMENT GENE IN STEM CELL GENOME.

Some stem cells, totipotent stem cells, can give rise to all tissues of an organism. Here, remarkably, a stem cell can give rise to a whole mouse!

ONLY CELLS WITH DRUG RESISTANCE GENE SURVIVE

BROTH WITH DRUG

CULTURE DISH WITH GROWTH-INHIBITING DRUG; ONLY KNOCK-OUT CELLS GROW.

By comparing knock-out and wild type mice, one can begin to understand how a gene contributes to the phenotype of the animal.

INJECT STEM CELLS INTO EMBRYOS, WHICH ARE IMPLANTED IN UTERUS OF SURROGATE MOTHER.

GROW MOUSE EMBRYOS FROM KNOCK-OUT CELLS

KNOCK-OUT MOUSE LACKING PIGMENT GENE

Cloning the Organism – The History

In 1928, Hans Spemann performed the first 'nuclear transfer' experiment. Using a baby's hair he eased the nucleus from the cell of an embryo and squeezed this nucleus into an enucleated cell from a younger embryo.

it's the CLONE SHOW, folks!

An identical embryo was produced.

SPEMANN

171

An important question remained unanswered. Could an adult cell – a mature liver cell, heart muscle cell or a skin cell – be reprogrammed to develop into an entire organism? Could the genetic mechanism be 'rewound' to start all over again?

In 1938 Spemann, who had become the director of the Kaiser Wilhelm Institute of Biology in Berlin, proposed a 'fantastic experiment': remove the nucleus from the adult cell of an animal and put it in an enucleated egg. Spemann wanted to prove once and for all that an adult cell could be reprogrammed to re-create a copy of the animal from which it came.

Proof came in 1962 when John Gurdon in Oxford used a fully differentiated intestinal cell nucleus to clone a frog. In 1996 the first mammalian clone – Dolly, a lamb – was born, cloned from an adult cell (in this case a mammary cell from the udder). Spemann was proved right. Since then numerous animals – among them pigs, cats, cows, goats, rabbits, horses – have been successfully cloned. Yet the cloned animals are often abnormal.

GURDON

FROM THE NUCLEUS OF AN INTESTINAL CELL TO A WHOLE FROG

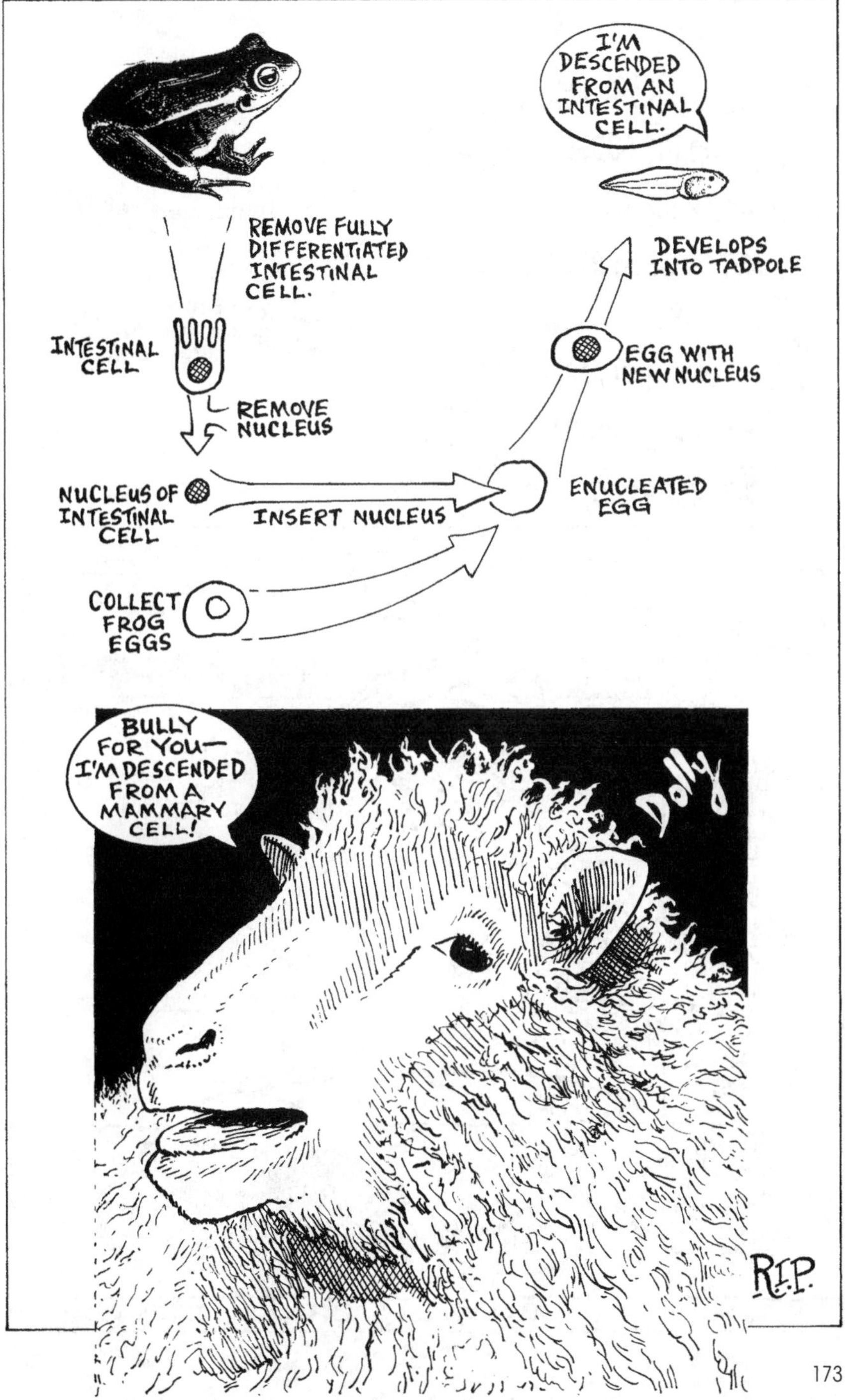

MAKING TISSUES FROM STEM CELLS

Stem cells:
1) have not yet specialized their functions;
2) retain the capacity to differentiate into specialized cell types;
3) renew themselves;
4) can be used to treat disease.

Stem cells may make it possible to grow new organs and tissues in the laboratory and to replace tissues and organs that are diseased, such as in heart, kidney or liver failure.

The rapid progress in stem cell research has also created a debate over the ethics of cloning. And new issues are likely to emerge, such as cloning of individuals or even the creation of variant life-forms.

In 2006, Kazutoshi Takahashi and Shinya Yamanaka working in Kyoto were able to get mature mouse cells to return ("revert") to an embryonic state, opening the possibility of creating clones without gene transfer.

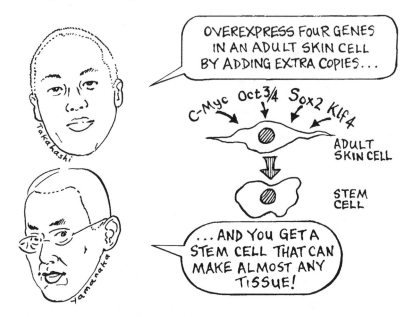

OVEREXPRESS FOUR GENES IN AN ADULT SKIN CELL BY ADDING EXTRA COPIES...

C-Myc Oct3/4 Sox2 Klf4

ADULT SKIN CELL

STEM CELL

...AND YOU GET A STEM CELL THAT CAN MAKE ALMOST ANY TISSUE!

The rapid progress in stem cell research has raised the possibility of generating new nerve cells, heart tissue, liver, bone or other organs to replace diseased or accidentally damaged tissues. Stem cells are also an essential part of contemporary research into the mechanisms of human development and disease.

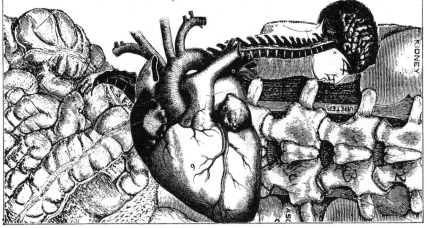

THE IMPACT OF NEW GENETICS ON MEDICAL RESEARCH

WHAT EXACTLY ARE WE DOIN', WILFRED?

TETRACYCLIN'

In seeking to create animal models of human diseases, scientists have engineered genes that can be turned on or off by drugs such as tetracycline (tet). By studying mice in which the ras oncogene was turned on by tetracycline, Ron DePinho at Harvard discovered that this oncogene both starts tumors and keeps them growing. Tumors formed when tetracycline was included in the diet and the tumors regressed when it was removed.

ROAD TO CANCER RESEARCH

DePinho

FEEDING THE MOUSE TETRACYCLINE TURNS ON THE RAS ONCOGENE AND CAUSES A TUMOR.

CHOW WITH TETRA-CYCLINE

These results suggest that tumors might be treated by turning off specific oncogenes.

myc

ras

brca

Oncogenes arise from normal genes by mutation (ras, brca) or mutation plus overexpression (myc). Brca mutation increases breast cancer risk.

Oncogenes

Most tumors require the cooperation of several oncogenes, the combination of which may vary from tumor to tumor. By establishing the profile of oncogene expression of a patient's tumor, an individualized therapy of oncogene inhibition may be chosen.

THOSE ONCOGENES AREN'T NEARLY INHIBITED ENOUGH.

In chronic myelogenous leukemia (CML), gene mutation and rearrangement form an oncogene that encodes an oncogenic enzyme, the Bcr-Abl protein, which attaches phosphates to other cancer proteins, making cells divide and form tumors.

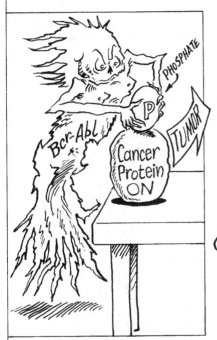

Gleevec, one of the first designer drugs, blocks Bcr-Abl's phosphate-attaching activity and brings about remission in CML patients. However, over time, the Bcr-Abl oncogene may mutate again, so that its protein is no longer inactivated by Gleevec, leading to a relapse. Scientists are designing successors to Gleevec that target the Gleevec-resistant forms of Bcr-Abl. In the future, by determining the identity of a tumor's oncogene, individualized therapy will become possible.

DIAGNOSIS

DNA can be used to diagnose disease. Huntington's disease is a progressive and incurable hereditary disease of the central nervous system. The first symptoms, uncontrolled movements, clumsiness, inability to concentrate and depression, usually appear when diseased individuals are in their 30s to 50s. The disease results from CAG triplet nucleotide repeats – which encode glutamines – in the DNA of the Huntingtin gene.

A Huntingtin gene with fewer than 26 CAG repeats is normal, but if the number is large, greater than 39 repeats, the Huntingtin protein is toxic and fatal. The availability of a definitive diagnosis, determining triplet repeat number by sequencing the Huntingtin gene, raises a quandary for individuals whose parents have the disease whether to get tested, because they have a 50-50 chance of inheriting the toxic mutant.

THE FAMED FOLK SINGER, WOODY GUTHRIE, DIED ON OCTOBER 3, 1967 OF HUNTINGTON'S DISEASE. HE WAS 55 YEARS OLD.

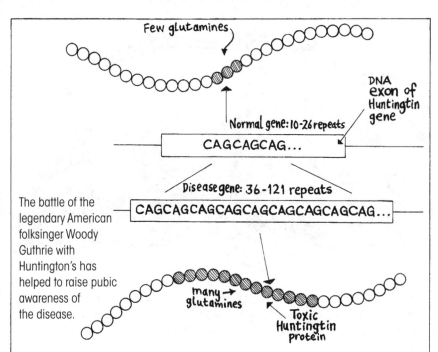

Few glutamines

DNA exon of Huntingtin gene

Normal gene: 10-26 repeats

CAGCAGCAG...

The battle of the legendary American folksinger Woody Guthrie with Huntington's has helped to raise pubic awareness of the disease.

Disease gene: 36-121 repeats

CAGCAGCAGCAGCAGCAGCAGCAGCAG...

many → glutamines

Toxic Huntingtin protein

Glutamine repeats underlie several other neurodegenerative diseases in which muscle control is lost. In the disease Fragile X Syndrome, the triplet CGG is repeated. The CpG sequence within this triplet can, as we have seen (see page 158), be methylated, causing a constriction in the chromosome that makes the chromosome fragile. It also silences genes, leading to mental retardation.

Guthrie

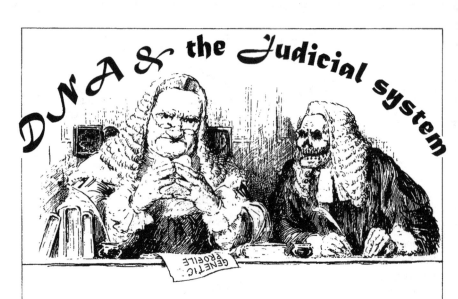

DNA plays a central role in deciding guilt and innocence in courtrooms around the world. Human DNA contains short non-coding DNA sequences (9 to 80 bases long) that may be repeated up to thirty times, called Variable Number of Tandem Repeat (VNTR) sequences.

VNTRs are found many thousands of times in human DNA, and can be detected in trace amounts of DNA such as from hair roots, or in biological fluids. DNA is collected with a swab, amplified by PCR (see page 185), and cut into small specific fragments using restriction enzymes (see page 114). The DNA fragments are applied to a gel and separated by size when an electric current is passed through the gel. Small fragments moving rapidly and large ones, slowly (see below).

The DNA is transferred to a membrane and probed (see Glossary, under **probe**) with radioactive DNA, using a method called DNA hybridization, to visualize the VNTR bands and reveal their sizes. The pattern of bands from the crime scene DNA is compared with DNA bands from the victim and the suspects. Because the number of VNTRs that lie between two EcoR1 sites will vary from person to person, the pattern of the VNTRs can identify an individual, and be used in the courtroom like a fingerprint.

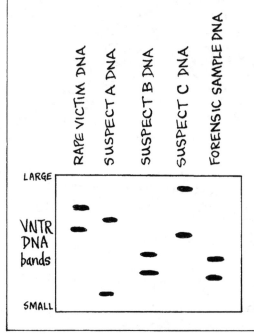

An example of a gel used to compare VNTR DNA fragments from four people, three suspects (A, B and C) and a rape victim, with the fragments from DNA from semen obtained from the victim (forensic sample). The DNAs were digested with a restriction enzyme to create the VNTR fragments. These were separated on a gel according to size and visualized by autoradiography. The VNTR bands from each person were different. One matched the VNTR bands from the forensic sample, identifying suspect B as the rapist.

Nonetheless, although the chance may be small (only one in a million), a person whose ethnicity is similar or who is a relative would have a significant chance of sharing the same VNTR patterns.

On the other hand, DNA can be used to prove an individual's innocence with virtual certainty. In the United States, the Innocence Project has successfully used DNA evidence to challenge the convictions of more than 200 individuals, many of whom had been incarcerated for over 15 years. Seventeen had been sentenced to death.

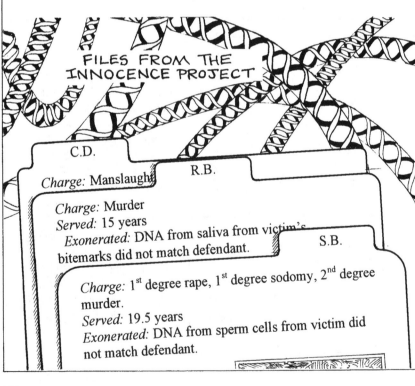

FILES FROM THE INNOCENCE PROJECT

C.D.

Charge: Manslaugh[t]

R.B.

Charge: Murder
Served: 15 years
Exonerated: DNA from saliva from victim's bitemarks did not match defendant.

S.B.

Charge: 1st degree rape, 1st degree sodomy, 2nd degree murder.
Served: 19.5 years
Exonerated: DNA from sperm cells from victim did not match defendant.

PCR

The polymerase chain reaction (PCR) replicates DNA outside a living organism. PCR's inventor, Kary Mullis, described its great power: "Beginning with a single molecule of the genetic material DNA, the PCR can generate 100 billion similar molecules in an afternoon. The reaction is easy to execute. It requires no more than a test tube, a few simple reagents, and a source of heat."

The DNA fragment to be replicated is mixed with short DNA primers, complementary to the ends of the DNA to be replicated. A heat-stable DNA polymerase, which synthesizes DNA copies, and the nucleotide precursors to the DNA itself are added. The tube is heated to unwind all of the DNA stands and then cooled, whereupon the primers bind to the ends of the fragment that is being replicated. The polymerase extends the primer, making one full double-stranded copy of each DNA fragment strand. The tube is heated again to unravel the strands and the process is repeated, doubling the quantity of the DNA. Each repetition of the reaction increases the copy number exponentially, producing many copies from as few as a single DNA molecule template. (Pshew . . . that was hard!)

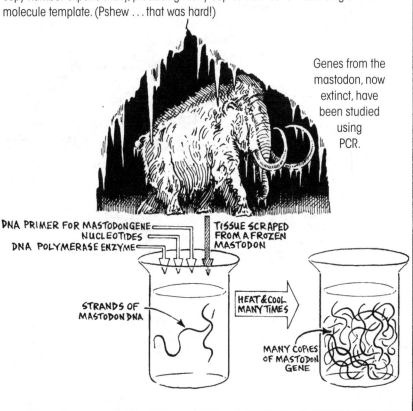

Genes from the mastodon, now extinct, have been studied using PCR.

DNA PRIMER FOR MASTODON GENE
NUCLEOTIDES
DNA POLYMERASE ENZYME

TISSUE SCRAPED FROM A FROZEN MASTODON

STRANDS OF MASTODON DNA

HEAT & COOL MANY TIMES

MANY COPIES OF MASTODON GENE

Because very small quantities of DNA are required, PCR can be applied to amplifying crime scene samples or even to ancient DNAs, such as DNA traces found in the flesh of 10,000-year-old mastodons, frozen in the tundra, or to DNA from mummies.

Michael Smith used PCR to mutate DNA by using a primer that differs at one residue from the wild type sequence. Site directed mutagenesis is a powerful tool for studying DNA and protein function.

DNA Machines

The structure of DNA is governed by base-pairing, with A binding to T and G to C in the double helix. The nucleotides can be viewed as snap together parts, which may be used to assemble complex structures, such as the truncated DNA octahedron constructed by the lab of Nadrian Seeman. Seeman is employing DNA in the new science of nanotechnology to construct machines on the molecular scale, such as nanorobots, perhaps the ultimate in miniaturization. Ultra-compact memory chips may one day be crafted from DNA, and store vast amounts of information in a device of molecular dimensions.

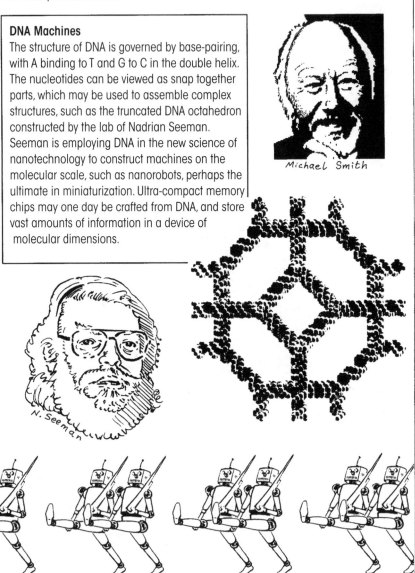

Michael Smith

N. Seeman

DNA is Nature's magical molecule. How did our understanding of DNA shake the world...?

read on...

187

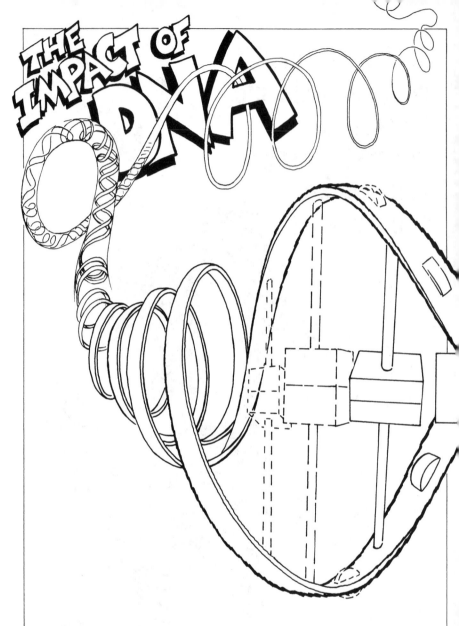

THE IMPACT OF DNA

Cloning not only taught us much about gene structure, but also captured the public imagination — and later its fears. The first public debate over the benefits and hazards of genetic engineering came in 1973 when scientists active in gene research considered the implications of cloning for society.

Biologists recognized that the most likely effect of reorganizing DNA would be to make it, simply stated, non-functional. Genes have evolved through eons of selection, mutational change and natural trial.

Transferring human genes into bacteria would therefore probably disrupt their function rather than create a hazard in the biosphere.

However, concern remained that a novel recombinant could have undesirable properties. For example, a bacterium synthesizing insulin in the gut of a human might imbalance the person's sugar metabolism.

With a call for a moratorium in 1974, scientists voluntarily deferred certain classes of cloning experiments. Later strict guidelines for conducting cloning were adopted by the scientific community.

One critic questioned:

What if an ant crawled from the bio-containment facility with a perilous *E. coli* perched on its thorax?

Opposing scientists grappled in public debate.

Most poignantly, Erwin Chargaff and James Watson, who once exchanged scornful and mistrusting glances in the Cavendish Laboratory, assaulted each other's positions on recombinant DNA.

Said Chargaff:

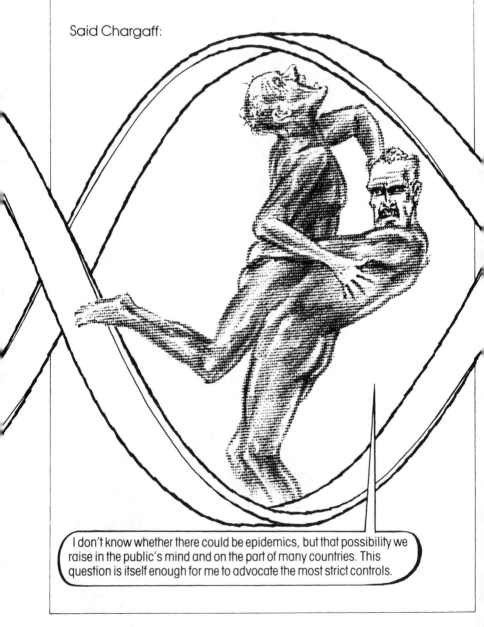

I don't know whether there could be epidemics, but that possibility we raise in the public's mind and on the part of many countries. This question is itself enough for me to advocate the most strict controls.

Said Watson:

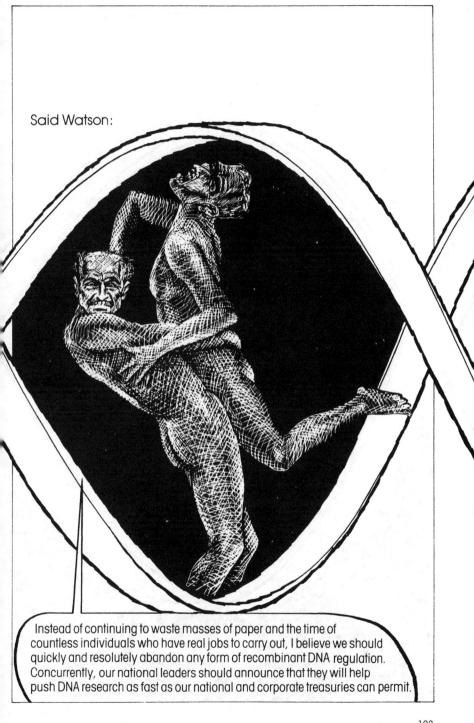

Instead of continuing to waste masses of paper and the time of countless individuals who have real jobs to carry out, I believe we should quickly and resolutely abandon any form of recombinant DNA regulation. Concurrently, our national leaders should announce that they will help push DNA research as fast as our national and corporate treasuries can permit.

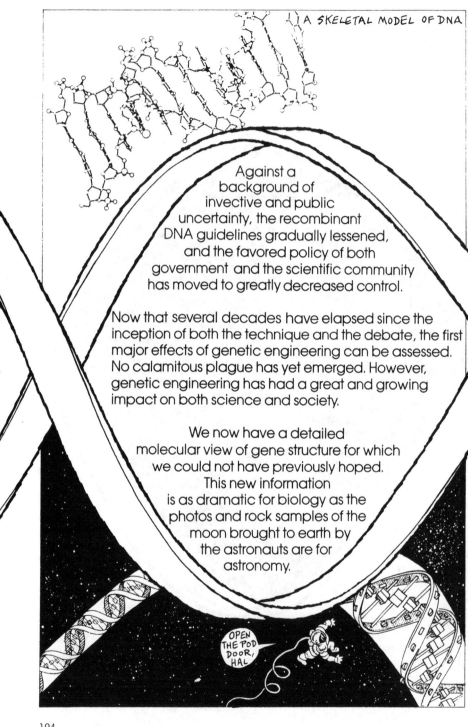

Against a background of invective and public uncertainty, the recombinant DNA guidelines gradually lessened, and the favored policy of both government and the scientific community has moved to greatly decreased control.

Now that several decades have elapsed since the inception of both the technique and the debate, the first major effects of genetic engineering can be assessed. No calamitous plague has yet emerged. However, genetic engineering has had a great and growing impact on both science and society.

We now have a detailed molecular view of gene structure for which we could not have previously hoped. This new information is as dramatic for biology as the photos and rock samples of the moon brought to earth by the astronauts are for astronomy.

OPEN THE POD DOOR, HAL

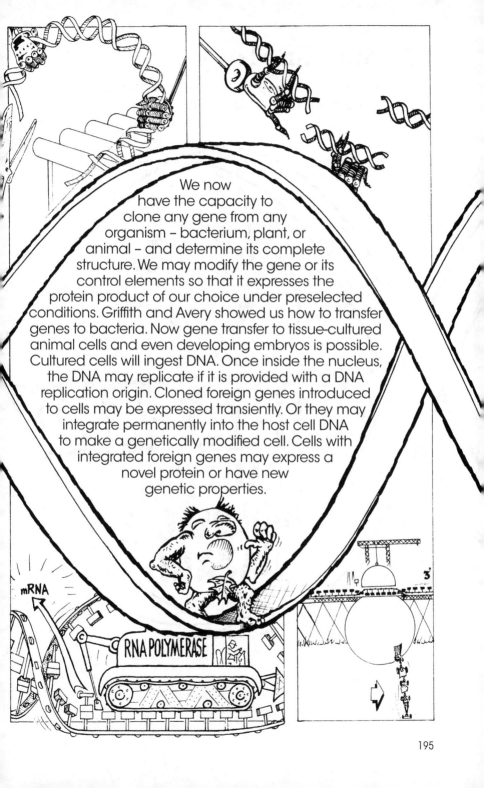

We now have the capacity to clone any gene from any organism – bacterium, plant, or animal – and determine its complete structure. We may modify the gene or its control elements so that it expresses the protein product of our choice under preselected conditions. Griffith and Avery showed us how to transfer genes to bacteria. Now gene transfer to tissue-cultured animal cells and even developing embryos is possible. Cultured cells will ingest DNA. Once inside the nucleus, the DNA may replicate if it is provided with a DNA replication origin. Cloned foreign genes introduced to cells may be expressed transiently. Or they may integrate permanently into the host cell DNA to make a genetically modified cell. Cells with integrated foreign genes may express a novel protein or have new genetic properties.

mRNA

RNA POLYMERASE

3'

195

Biotechnology

With public fears of genetic engineering receding, scientists considered applying cloning to biotechnology.

Biotechnology is the commercialization of biology and genetics. It is the application of new genetic technology to practical medical and industrial problems. Biotechnology arose in San Francisco, not far from Silicon Valley, the birthplace of the transistor, micro-chip, and computer industries.

The first projects were to transfer the genes for medically important proteins such as growth hormone or insulin to bacteria where these proteins might be cheaply produced in abundance.

The giddy stage of investment passed and many ask how, realistically, biotechnology might profit society. Here are some possiblities. Biotechnology promises to:

1. Provide diagnostic reagents for detecting genetic diseases such as Down syndrome, sickle cell anemia, or even somatic genetic diseases such as cancer.

2. Produce vaccines against diseases of livestock and (with government approval) humans. Some vaccines, such as for malarial parasites – which cleverly avoid immune detection – might not be feasible through other approaches.

3. Produce hormones, blood clotting factors, insulin or other protein pharmaceuticals such as interferon. In the future new complex and specifically targeted "protein drugs" may be possible with gene cloning.

4. Produce industrial chemicals such as the sweetener fructose. Woodchips might be converted to sugar or to synthetic fuels such as gasohol.

5. Genetically modify plants for the mass production of chemicals and proteins and novel nutrients fueled by cheap photosynthetic energy. Bypass natural barriers for gene transfer and overcome slow breeding times to make plants disease-resistant and viable in soils previously unsuitable for agriculture.

6. Develop new energy sources and animal feed stocks through waste and biomass recycling.

7. Salvage precious metals, develop new bio-mining techniques for the recovery of ore metals. Control pollution.

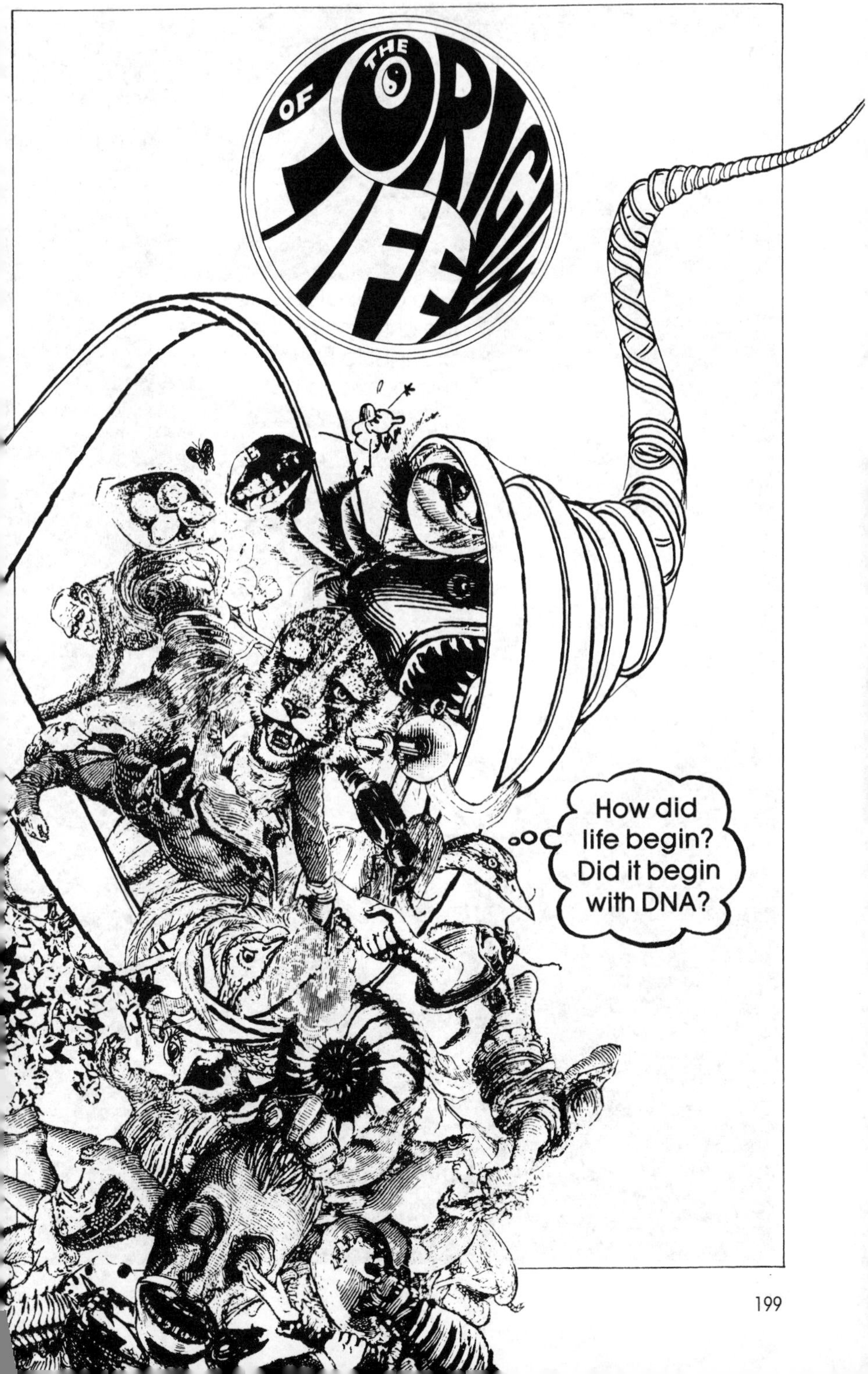
OF THE
How did life begin? Did it begin with DNA?

In the nineteenth century the Swedish physicist Gustav Arrhenius suggested the theory of **panspermia**.

"Life originated elsewhere in the universe and has been brought to Earth by microorganisms."

A few years ago, Leslie Orgel and Francis Crick resuscitated the theory:

But what if life did begin on earth. How could it have started? In 1924 the Russian biochemist A. I. Oparin published a monograph (little noticed at the time):

That early atmosphere of the Earth probably contained methane gas, ammonia and water, but **no** oxygen. Ultraviolet light from the sun, electrical storms, and volcanic activity would have led to the formation of a "prebiotic" soup, a mixture of organic materials that would become the building blocks of primitive life.

WHOOPS! I DON'T THINK WE EVOLVE FOR A FEW HUNDRED MILLION YEARS YET!

In the 1950's, Harold C. Urey and Stanley L. Miller...

...mixed ammonia, methane, hydrogen, and water together in a large flask and to simulate the electrical storms subjected the mixture to periodic electrical discharges.

Within days amino acids began to accumulate in the apparatus!

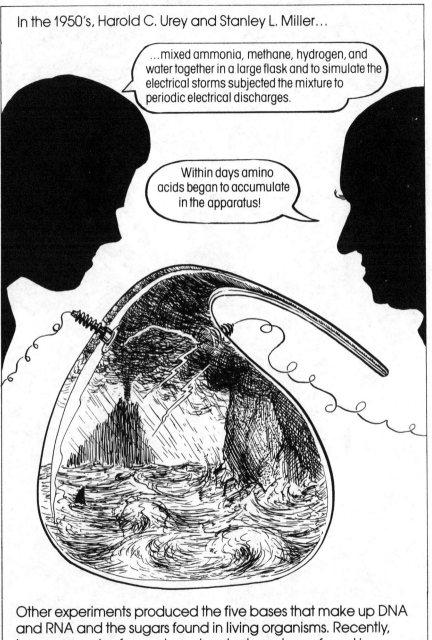

Other experiments produced the five bases that make up DNA and RNA and the sugars found in living organisms. Recently, large amounts of organic molecules have been found in interstellar space. And meteorites from outer space contain amino acids in considerable quantities.

Prebiotic soups could well be forming in many parts of the universe.

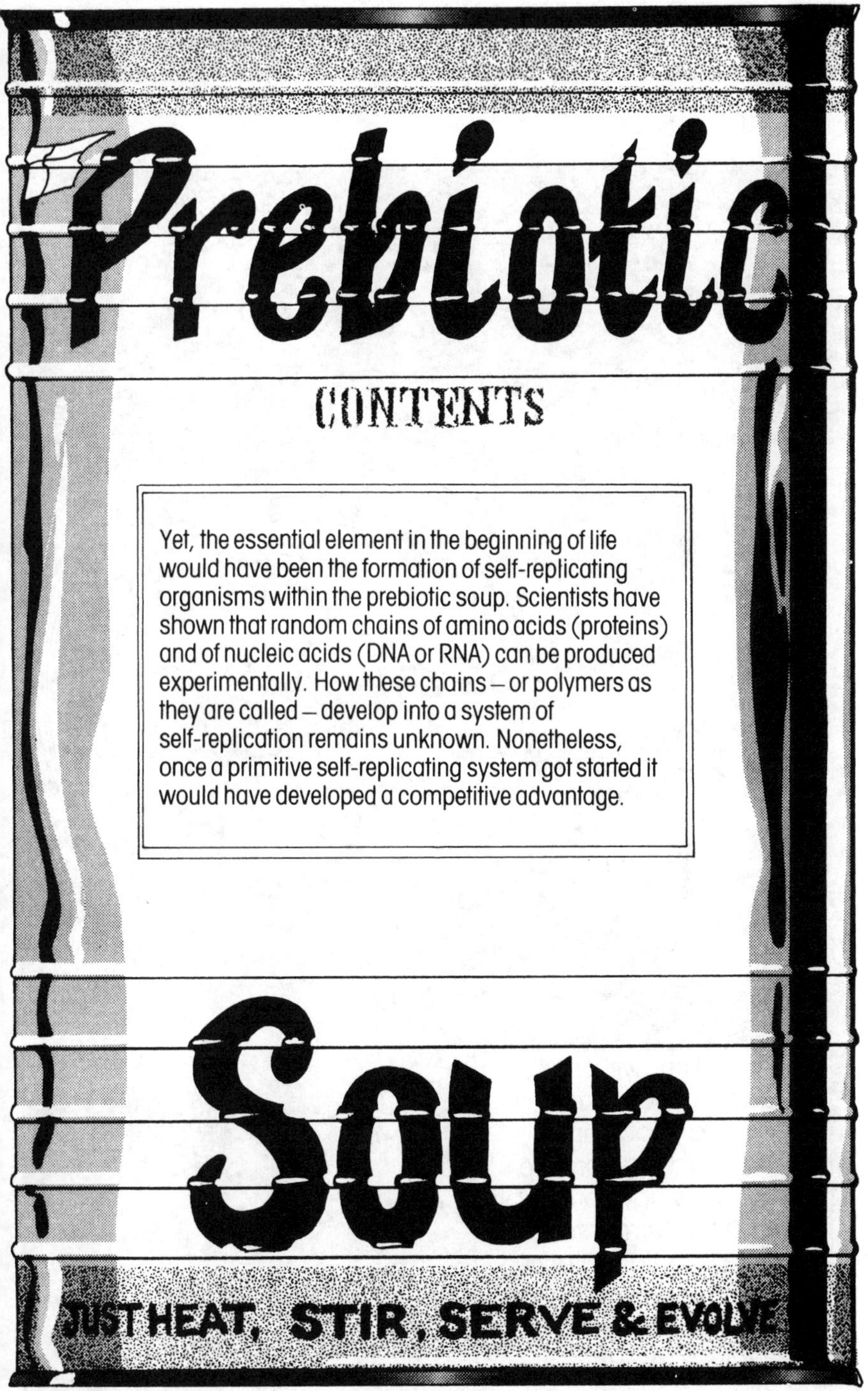
Prebiotic
CONTENTS
Yet, the essential element in the beginning of life would have been the formation of self-replicating organisms within the prebiotic soup. Scientists have shown that random chains of amino acids (proteins) and of nucleic acids (DNA or RNA) can be produced experimentally. How these chains – or polymers as they are called – develop into a system of self-replication remains unknown. Nonetheless, once a primitive self-replicating system got started it would have developed a competitive advantage.
Soup
JUST HEAT, STIR, SERVE & EVOLVE

Random aggregations of molecules and the **selection** of the more successful duplicating processes were, most probably, the driving forces in the formation of life.

The oldest known primitive organisms, found in sedimentary rocks in Australia dating from 3.6 billion years ago, probably lived off the fermentation of various organic materials. Photosynthetic processes evolved and they began producing oxygen.

For one hundred million years the oxygen produced by these organisms reacted with the iron in the oceans and precipitated the iron out as gigantic bars which form the major portion of today's iron reserves.

Only after the oceans had been 'rusted' did oxygen begin to fill the atmosphere.

The presence of oxygen forced many primitive organisms into the protective cover of oxygen-free environments.

Those cells that could tolerate an oxygen atmosphere evolved mechanisms for using the oxygen.

A GALACTIC VIEW OF DNA

There is an interesting side to the evolutionary process that is illuminated by astronomy. The living organisms we now see all have their structure based upon the element carbon. Most biochemists believe no other basis is possible for life. But where does carbon come from? Carbon originates in the centre of stars where at temperatures of millions of degrees it is 'cooked' from simple protons and neutrons. When the stars reach the end of their lives they explode and disperse carbon into space and on to the surface of planets and meteorites. However, the time needed to make carbon and other heavier elements, like nitrogen and oxygen, by this stellar alchemy is very long: nearly a billion years.Only after this immense period of time will the building blocks of life be available in the universe, and only then can biochemistry take over.

So, life is only possible in a universe that is at least a billion years old. Remarkably, because the universe is in a state of expansion, this also means that life can only arise in a universe that is at least a billion light years in size. The vastness of the universe is inextricably bound up with the existence of life within.

John D. Barrow

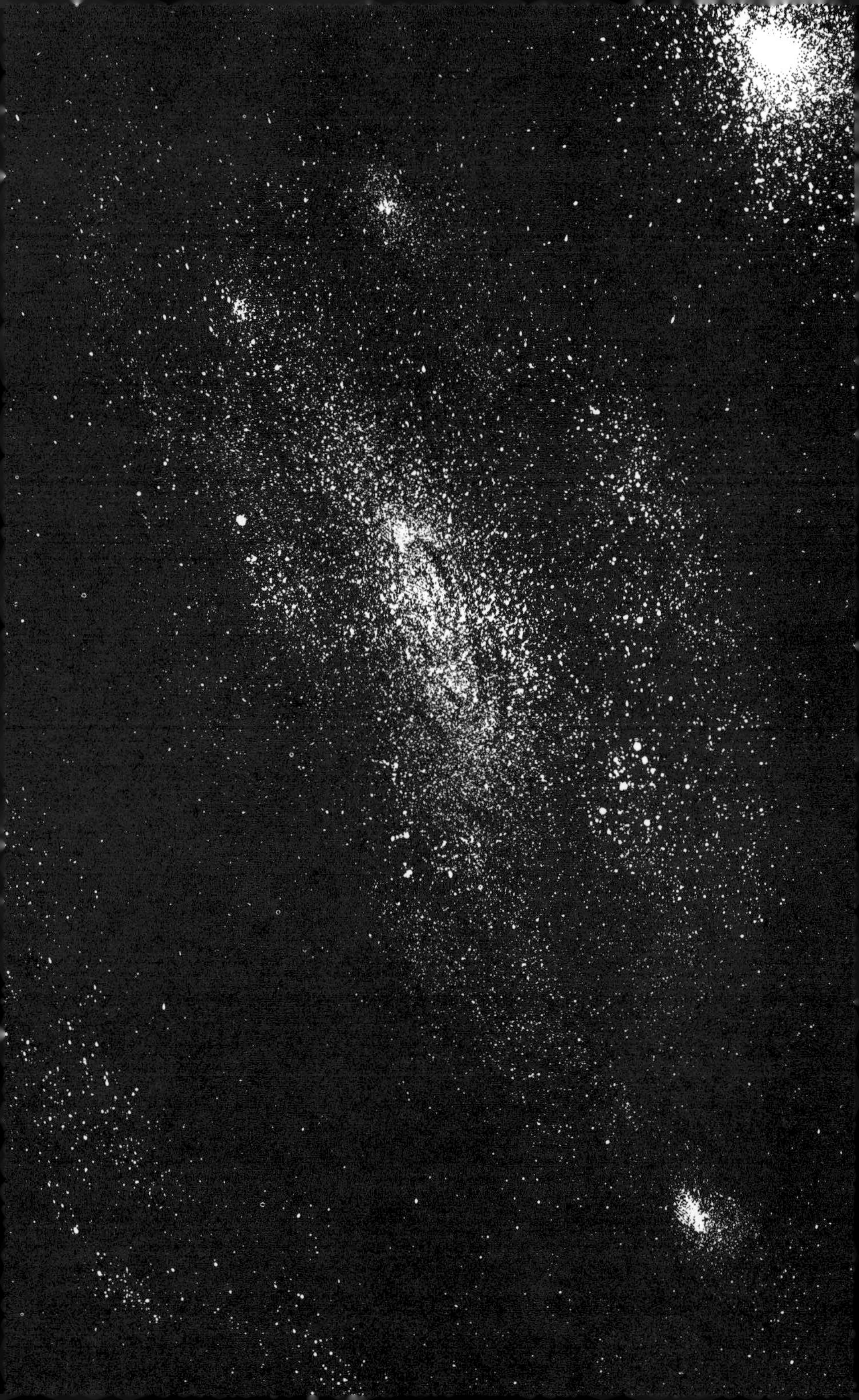

About 1.4 billion years ago the first cells with nuclei –
the eukaryotic cells (eu = true; karyote = kernel) –
appeared. Advanced sexual reproduction was
now possible with a consequent
more rapid pace to evolution.
By one billion years ago
multicellular organisms
had evolved.

The history of the earliest life-forms shows how changes in the environment created new selective pressures, giving rise to new life-forms.

But what is natural selection selecting? And does our knowledge of the structure of DNA give us any insights into the possible molecular mechanisms?

Variations appear to be randomly produced. Many do not help the organism adapt to its environment. (As Stephen Jay Gould and Richard Lewontin have said: the "male tyrannosaurs may have used their diminutive front legs to titillate female partners, but this will not explain **why** they got so small.")

—BUT THANK THE LORD THEY DID!

Natural selection does not foresee the future.

TINKERING

Adaptation is not like solving an engineering problem. The jet engine did not **evolve** from the combustion engine, but was built from scratch. Biological organisms must somehow incorporate what is already there into the new organism.

François Jacob:

> Evolution proceeds like a tinkerer who, during millions of years, has slowly modified his products, retouching, cutting, lengthening, using all opportunities to transform and create.

I TINKER THEREFORE I AM

Because the globin gene – discussed earlier – was duplicated, an extra copy was available for tinkering. Mutation and natural selection could then create globin diversity.

SELFISH GENES

In 1976 Richard Dawkins published his book **The Selfish Gene**, creating a considerable stir throughout the scientific and even philosophical communities. Dawkins argued that selection is at the gene level.

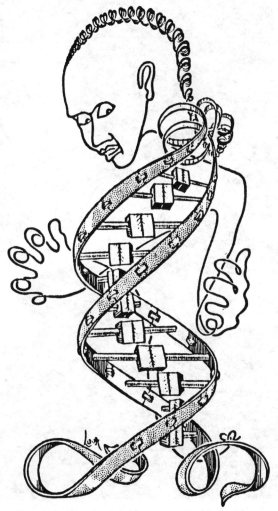

The aim of a gene, he said, is to survive from one generation to the next and it **uses** the bodies of living organisms.
Human beings are simply survival machines for DNA.

Then in 1980 Francis Crick and Leslie Orgel presented the ultimate argument for self-centered molecules: **selfish DNA**. Some DNA exists, they said, not because of any benefits it might bring to an organism, but because that DNA is what is being selected in evolution.

Not all DNA is "selfish" and the organism lets the selfish DNA exist as long as it doesn't get too much in the way, because it would take too great an effort for the organism to get rid of the "junk" DNA.

The selfish gene (and selfish DNA) idea, with its metaphorical image of living things being manipulated by DNA, had its genesis in a paper written by the English population geneticist D.W. Hamilton in 1964. He tackled a problem that had puzzled Darwin:

The evolution of sterility in certain species of insects (which include ants, bees, and wasps).

Hamilton noted that the females of these species have pairs of each chromosome: Two sets.

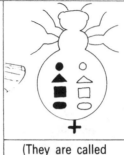

(They are called diploid.)

Males however have one of each chromosome: One set.

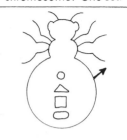

(They are called haploid.)

To parent a daughter:

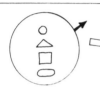

The father gives his entire set – *the same for each daughter.*

The mother also gives one set, but she draws this set by taking some chromosomes from her first set and some from the second. . .

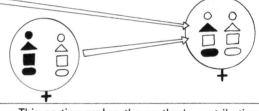

. . . This sorting makes the mother's contribution *different for each daughter.*

8 SISTERS. . . SISTERS. . . NEVER WERE THERE SUCH DEVOTED SISTERS!

When we do the sums, sisters are more closely related to each other than to their mothers, or their offspring.

Therefore **if** genes are selfish, sisters (even sterile ones) are better off helping each other than their offspring if they want their own genes to survive. This is exactly the way these insects behave.

Of course, this doesn't prove that genes **are** selfish.

Curious, isn't it, though, that the ants and bees behave just as you would expect them to if they were being controlled by selfish genes?

It would be foolish to draw conclusions about human behavior based on analogies with insect behavior (which is largely programmed by their genetic make-up). Human behavior is determined by genetic factors against a powerful background of cultural and moral beliefs and relationships. An ant could never avenge an ancestor's death, believe in God ... or discover DNA!

Evolving Evolution – Sources of Darwinian Theory

DNA is a marvel that has transformed society. However, to understand the full impact of DNA, we must appreciate Darwinian theory and its sources. Charles Darwin wrote *On the Origin of Species* in 1859, the keystone of modern evolutionary theory. As the English theoretical biologist John Maynard Smith wrote, "No other writer had such a profound effect on the way we see ourselves, and no other brought about so great an extension in the range of subjects which we regard as explicable by scientific theory." As we shall see, Darwin argued that all existing organisms come from one or a few common ancestors, and that evolutionary change arises from the natural selection of variant life-forms. Darwinian theory came out of a variety of economic, political, scientific and religious writings about the origins and nature of life.

"FREE MARKET COM-
PETITION BENEFITS
SOCIETY."

My economic theory of capitalism described in *The Wealth of Nations* makes the famous claim that society consists of selfish individuals who in the pursuit of their own interests – enriching themselves – are also acting in society's best interest since society too is enriched. The free market is governed by a "hidden hand" - the laws of supply and demand – which are laws of nature not written by man.

Adam Smith- Economist

My *Natural Theology* uses the famous watchmaker analogy to assert that a divine intelligence is necessary to account for the complexity of life forms. Just as the complexity of a watch requires a watchmaker, the complexity of living forms requires an intelligent designer.

"GOD DESIGNED
THE WORLD."

William Paley- Bishop & Philosopher

"NATURE EVOLVES
GRADUALLY."

My theory of gradualism espoused in *Principles of Geology* says that geological changes occur gradually over extended periods of time. For example, the Grand Canyon, which was carved out gradually by the Colorado River over hundreds of millions of years.

Charles Lyell- Geologist

My treatise *An Essay on the Principle of Population* claims that populations grow exponentially (2, 4, 16…), while the food supply grows arithmetically (1, 2, 3…), leading to a struggle for the scarce resources – war famine and political unrest.

"POPULATION
OUTSTRIPS SUPPLY."

Thomas Malthus- Political Economist

Darwin transformed these late-18th- and early-19th-century economic, geological, religious, and political ideas into a radically new biological theory. Smith's theory became natural selection guided by a hidden hand; Lyell's theory became Darwin's theory of the accumulation of small changes (mutations) over huge amounts of time, known as gradualism.

Paley's watchmaker raised the question of the origin of precision in an organism's match to its environment. In the theory of Intelligent Design, precision comes from an engineer's designing a device, the watch, which has a specific goal to tell time. A watch is precise, according to Paley, because a watchmaker has designed it so that it is an accurate time-keeper. However, a living form differs greatly from a mechanical device. Living forms must have the capacity to transform into newer forms should the environment change. Since the environment's changes are unpredictable, for an intelligent designer to be capable of designing new life forms with precision, the designer would have to be a soothsayer, able to predict the future. Darwin found a way of creating a much greater precision, through random variation and selection, a process that might be called unintelligent design, or lack of any design at all.

Contrary to popular belief, randomness coupled with selection establishes the possibility of the most precise fit to the environment.

In sum, Darwin's theory of evolution is based on three ideas: natural selection, heredity and variation. Small random changes – variations – occur in organisms from one generation to the next. The variant traits that are selected are passed on, through reproduction, to the next generation of organisms. "Natural selection," Darwin wrote, "acts solely by accumulating successive, favorable variations." Evolution in the Darwinian view was gradual: "it can act only by short and slow steps." All living organisms, Darwinian theory claimed, are descended from one or a few common ancestors.

Neither Darwin nor any of his contemporaries knew anything about heredity. Yet as we have seen, the new science of "genetics," the idea of "genes" transmitting specific traits, such as hair color, from one generation to the next, began in the first decade of the 20th century.

By the 1940s, though the structure of the gene was still unknown, scientists had integrated the idea of the gene into Darwinian theory. They now explained evolution as the consequence of small random changes in genes. This recasting of Darwinian theory was called the Modern Synthesis, following the 1942 publication of Julian Huxley's book *Evolution: The Modern Synthesis*. This neo-Darwinian theory corrected Darwin's failure to explain the mechanism of inheritance. Embryology was not mentioned.

The neo-Darwinian view appeared to be spectacularly confirmed when the double helix was discovered in 1953, showing how genes composed of DNA transmitted hereditary characteristics.

Not only did the structure of DNA suggest a mechanism for gene replication, it also made apparent how variations arising from random changes were possible and could be inherited through changes in the base sequence of a gene. This idea of small random mutations in the base sequences of genes appeared to confirm Darwin's view, already mentioned, that nature "can act only by short and slow steps. Hence, the canon *Natura non facit saltum*, nature doesn't make sudden jumps. Evolution is gradual. The standard view, then, was that variation and selection could account for how the simple organisms of early life evolved into the complex forms of the contemporary biological world. It was assumed that as changes accumulated, there would be less and less similarity of genetic sequences from one species to another; and more advanced species would have many more genes. Worms would have few, if any, genes similar to those of fish, mice or human beings.

The Big
Gene Bet
of the Last
Millennium
(1999 AD)

ANTENNAPEDIA

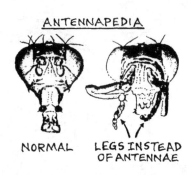

NORMAL   LEGS INSTEAD
          OF ANTENNAE

antenna replaced by a leg. These and other mutant forms revealed a set of eight genes in the fruit fly that controlled much of its embryonic development – the shape of its body and the distribution of the attached appendages. Very similar genes were subsequently found in worms, flies, fish, mice, and humans.

The Twists & Turns of History:
The Problems with Gradualism

In 1894, the English biologist William Bateson challenged Darwin's view that evolution was gradual. He published *Materials for the Study of Variation*, a catalog of abnormalities in insects and animals in which one body part was replaced with another. He called these abnormalities *homeotic* transformations. Among the forms he described was a mutant fly with a leg instead of an antenna on its head, and mutant frogs and humans with extra vertebrae.

These discoveries came as a great surprise to scientists, since the belief in small mutational changes in DNA molecules over hundreds of millions of years made the preservation of whole genes over long periods of time highly unlikely. Furthermore, the discovery that the same genes existed throughout the animal world, in fish, snakes, apes and human beings, wasn't anticipated. It was thought that each animal had evolved its own unique set of genes over millions of years and that this explained the diversity of living forms. Surely humans could not have the same genes as worms.

Beginning in the late 1970s, first Edward Lewis at Harvard and then Christiane Nüsslein-Volhard in Germany and Eric Wieschaus at Princeton began a systematic study of mutant flies, flies with four rather than two wings, or with an

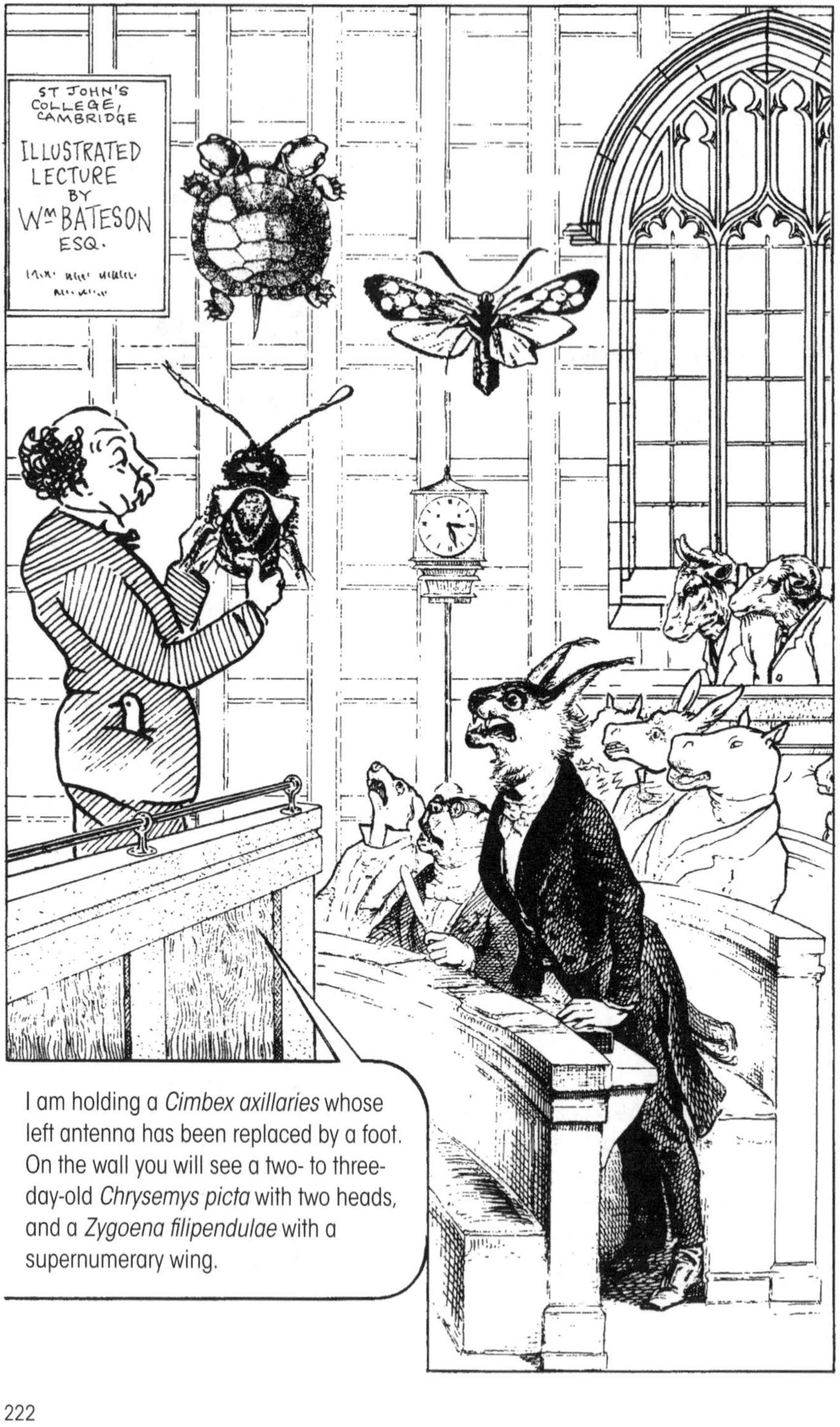
ST JOHN'S COLLEGE, CAMBRIDGE
ILLUSTRATED LECTURE BY Wm BATESON ESQ.
I am holding a *Cimbex axillaries* whose left antenna has been replaced by a foot. On the wall you will see a two- to three-day-old *Chrysemys picta* with two heads, and a *Zygoena filipendulae* with a supernumerary wing.

In 1983, Walter Gehring's laboratory in Basel found a short stretch of DNA (180 bases in length) that was virtually identical in all these newly discovered genes that, as we have said, controlled much of the embryonic development in living things from worms to humans. In homage to Bateson, the virtually identical sections were dubbed "homeoboxes," since they were present in genes that, when mutated, resulted in Bateson's "monsters" or "homeotic" transformations. Today these genes are called Hox genes, a combination of Bateson's "homeotic" and the more recent term "homeobox" (figure 1).

Even before the structure of DNA was known Barbara McClintock, working in her Cold Spring Harbor laboratory on Long Island, made a series of observations that would seriously challenge the idea that genes are simply a linear sequence of bases, though it would be decades before the true importance of her work would be realized.

McClintock spent most of her time studying the genetics of maize. She was particularly fascinated by the way the chromosomes of maize broke in specific places and then rejoined, creating mutations. She thought the genes were

Figure 1. The eight Hox genes of the fruit fly.

Hox genes have a critical part to play in determining the body plan of the fly. Top: the adult fly with the genes below it. Bottom: the embryo. Each gene regulates the development and identity of a specific region of the fly's body in the embryo and in the adult fly.

being turned on and off by "controlling elements" that could move about the chromosome – transposable elements, or "jumping genes." Formerly, it had been believed that mutations in genes were stable and would be passed on to future generations. But what McClintock was observing were mutations that were temporary and that were "undone" while the plant was growing. By 1948 – five years before the structure of the DNA molecule was worked out – McClintock had discovered that genetic elements changed places on the chromosomes. What was beginning to emerge – though no one fully grasped this at the time – was that genes are undergoing many changes during the lifetime of an organism.

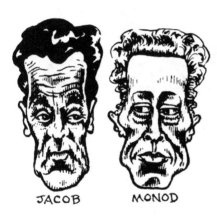

JACOB   MONOD

Scientists ignored or were unaware of McClintock's work. More than 10 years later, Monod and Jacob suggested a model of gene control that also depended on a control element or switch genes. Their control element, however, depended on a regulatory protein, the full significance of which only emerged with the discovery of the Hox genes.

The deeper meaning of the Monod-Jacob model of gene function became apparent, as we have already mentioned, when it was realized that part of the Monod-Jacob repressor molecule is strikingly similar to the part of the protein product of the Hox genes that is coded for by the homeobox, a segment of the Hox protein called the homeodomain. The similarity lies in the part of the repressor that binds to the DNA. Hox genes, then, like the Monod-Jacob repressor molecule, turn other genes on and off.

As we have seen (see page 163), another surprise occurred when the rough draft of the human genome was announced in 2001. As it turned out, human beings have far fewer genes than expected (about 25,000, not the 100,000 or more that had been predicted). Worms have about 14,000 genes and mice about as many genes as humans. The number of genes in a given species is not a measure of its complexity.

Why had biologists so overestimated the number of genes in the human genome? Why is it unnecessary for complex animals such as mammals to have ten times as many genes as worms?

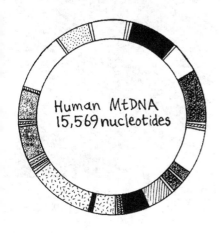

Human MtDNA
15,569 nucleotides

The discovery of Hox genes is only part of the story of how species evolved. The weakness of Darwinian theory – and one that has been seized upon by opponents of evolutionary theory – is its failure to explain how the gene determines the observable traits of the organism. From an evolutionary point of view, how can complex organs such as eyes, arms or wings evolve over long periods of time? What about the intermediary forms?

The Darwinian view was that early evolutionary forms of arms, legs, or wings might have initially served other purposes (insect legs, for example, might have evolved from gills their ancestors used for respiration). This is certainly important, but there must be other mechanisms at work here as well. If we return to the problem of the evolution of the eye, many questions arise: how is it possible for the different parts of an eye to evolve simultaneously – the lens, the iris, the retina, along with the blood vessels necessary for supplying oxygen and nutrition, as well as the nerves that must receive signals from the retina and send signals to the muscles of the eye? Could these precise nerve and vascular networks be created by gradual random changes over long periods of time, as Darwin claimed? Similarly, how can random mutations and natural selection create not only the necessary muscles and bone that make up the arm, but organize the blood supply and nerves so that, hundreds of thousands of years later, an animal evolves with functioning arms, legs and eyes? The Darwinian view that developing organs can serve different purposes at different times is incomplete at best.

Darwin thought that the direction evolution might take at any given time was purely random. In the neo-Darwinian view this meant that genetic variations were random. Hox genes create many different kinds of body plans. Yet, the individual parts of the body plan – contained within the different coordinates described above – can evolve independently of all the other parts of the animal. This independence means that mutations can occur that may or may not be beneficial, without being lethal for the developing embryo. In other words, while evolution is constrained by the body plan created by the Hox genes, this constraint gives nature a much greater freedom to experiment with variant forms through random mutations. If there were no body plan, most variations would be lethal and evolution would be much, much slower. Imagine we wanted to design new windows for airplanes that would improve the visibility for passengers, resist cabin pressure, and better insulate passengers from the cold. We would test the new window designs without changing their placement on the body of the plane. If we had to redesign the entire plane every time we changed the window design we would be much slower in developing new and improved planes. Similarly, Hox genes give the developing embryo a framework in which to experiment with new forms, such as wings and longer necks.

Virtually all animals today are bilateral, a form that appeared more than 500 million years ago. Hox genes are an essential element in the formation and development of bilateral morphologies throughout evolution. In 1909 in British Columbia, some of the oddest bilateral animals ever to have appeared on the face of the earth were discovered in a fossilized form, the Burgess Shale, dating from the earliest period of complex life forms (that is, the period before 500 million years ago). The morphologies in the Burgess Shale are part of what are called the Cambrian Explosion or Cambrian Big Bang.

Not all new evolutionary forms show obvious morphological similarities, yet they may have what Neil Shubin and co-workers have argued are "deep homologies": a "sharing of the genetic regulatory apparatus that is used to build morphologically and phylogenetically disparate features."[1]

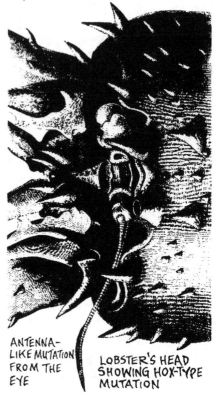

ANTENNA-LIKE MUTATION FROM THE EYE

LOBSTER'S HEAD SHOWING HOX-TYPE MUTATION

[Engraving from Bateson book]

"It is not possible to identify what is new in evolution without understanding the old. This is a reflection of the way evolution works, with some novelties being traceable as modifications of primitive conditions and others having origins that are much less obvious. As a result, the problems of novelty and homology have been deeply intertwined for the past

[1] Neil Shubin, Cliff Tabin, and Sean Carroll, "Deep Homology and the Origins of Evolutionary Novelty," *Nature* 457 (2009): 818–823.

century and a half.... One of the most important, and entirely unanticipated, insights of the past 15 years was the recognition of an ancient similarity of patterning mechanisms in diverse organisms, often among structures not thought to be homologous on morphological or phylogenetic grounds. In 1997, prompted by the remarkable extent of similarities in genetic regulation between organs as different as fly wings and tetrapod limbs, we suggested the term 'deep homology'."

Among the deep homologies of Shubin and co-workers are Hox genes, which are conserved throughout evolution and give rise to the "morphologically and phylogenetically disparate features" (such as wings and limbs).

The evolution of the eye is an example of a deep homology. The compound eye of the fruit fly is a collection of many independent light sensitive cell groups. The vertebrate eye has a single retina and many photoreceptor cells. Yet, vertebrate and fruit fly programs for constructing the eye depend on the same genes and transcription factors, demonstrating how

"[m]arkedly different eyes of long diverged phyla had more in common than was previously thought.... The unexpected finding that the homologous transcription factors Eyeless and PAX6 have crucial roles in the formation of the eyes of D. melanogaster (fruit fly) and vertebrates was the first indication that the markedly different eyes of long-diverged phyla had more in common than was previously thought. This discovery spurred comparisons of the detailed genetic circuitry underlying eye formation in diverse animals. It is now known that a small set of [proteins that regulate the transcription of genes] in D. melanogaster and their homologues in vertebrates, are widely used in the specification and formation of various types of animal eye.... This is a textbook example of deep homology: morphologically disparate organs whose formation (and evolution) depends on homologous genetic regulatory circuits" (figure 2).

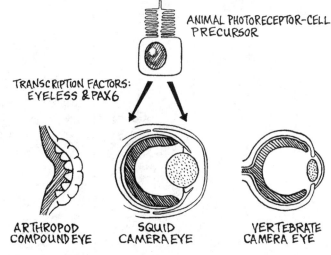

Figure 2. Deep homology in eye evolution. Although the arthropod, squid, and vertebrate eyes are morphologically very different, they all rely on the Eyeless and PAX6 transcription factors for their development. [Adapted from Shubin et al., "Deep Homology"]

Deep homologies are an example of what Mark Kirschner and John Gerhart call a "core process."[2] Core processes are metabolic and physiologic mechanisms that are conserved throughout evolution, making future developments more rapid by the reuse of these metabolic and physiologic pathways while allowing other features to diversify. The biochemical pathways which cells use to digest and metabolize nutrients are an example of a core process. These pathways were established at an early stage in evolution and are still used in human cells, worms and bacteria. Likewise the storage of genetic information in DNA and the mechanisms for translating that information in the synthesis of proteins are also core processes. Because of the deployment of these core processes

natural selection is presented with a variety of forms that are more likely to succeed than if there were no constraints on variation at all. However, should a new advantageous process arise, it can be incorporated into the functional repertoire of the organism, where it is carried forward over generations.

Another kind of core process that can create forms that are more likely to succeed is called "exploratory behavior." An example is the behavior of ants when foraging for food. Ants leave their nest and take random paths and secrete a chemical substance called a pheromone that leaves a scent. If an ant fails to find food it will eventually return to the nest,

---

[2] Mark W. Kirschner and John C. Gerhart, *The Plausibility of Life* (New Haven, Conn.: Yale University Press, 2005).

EXPLORATORY BEHAVIOR

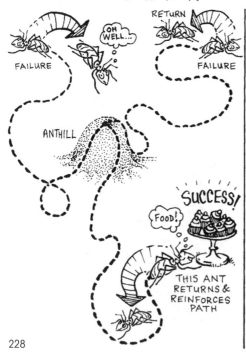

PATH REINFORCED BY SUCCESS

using the pheromones it has deposited to guide it back to the nest. However, an ant that finds food will deposit more pheromones as it returns to the nest. This will reinforce the scent of the trail that led to food and other ants will now follow the reinforced trail. Eventually, the ants will have established a detailed map of paths to food sources. An innocent observer might be fooled into thinking that the ants are using a map supplied by an intelligent designer of food distribution. However, what appears to be a carefully laid out mapping of pathways to food supplies is really just a consequence of a series of random search and selection (laying down of pheromones).

Other exploratory processes are important for the embryonic development of the vascular and nervous systems, and the guiding principles are similar to those of ant foraging: just as the ants randomly explore the terrain around their nest, capillary vessels sprout off the larger blood vessels and randomly explore the surrounding tissues for the signals coming from oxygen-deprived cells. And just as contact with food makes the ant reinforce the path that led to the food, the sprouting capillary vessels establish permanent contacts whenever they encounter tissue with oxygen-deprived cells. Similarly, fine nerve endings grow randomly, establishing stable nerve-muscle connections whenever they receive electrical and chemical signals coming from muscle.

Hence the evolution of organs such as the eye or the hand, with apparently well-designed and integrated nervous and vascular systems, does not require a global architectural plan with predetermined paths and wirings. Darwin's view that small simultaneous changes would give rise to organs as complex as the eye is in principle true, but in need of modification. It is the very constraints created by the Hox genes and the core processes (the exploratory behavior of capillaries and nerve endings) that permit complex designs to emerge over a relatively short period of time from a biological point of view (hundreds of thousands of years, or perhaps even less). Some genetic alteration is still necessary, of course, if the changes are to be passed on from one generation to another. But the genetic alterations are considerably simpler and fewer in number than we might have formerly imagined.

Watching Evolution in Action: Changes in the Stickleback Fish

Morphological variations take place in the bony armor of stickleback fish, for example those living in salt water as opposed to those living in fresh water. These changes could result from a need for protective armor by fish living in the ocean. The armor is a handicap for

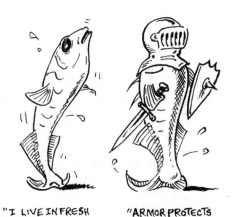

"I LIVE IN FRESH WATER — NO ARMOR, PLEASE."

"ARMOR PROTECTS THE MARINE STICKLEBACK FROM BIG PREDATORS."

those living in fresh water streams where invertebrate predators may attach themselves to the armor, and eat the fish.

Changes in the expression of a single gene, the Pitx1 gene, cause the changes in morphology.

HINDLIMB REDUCTION

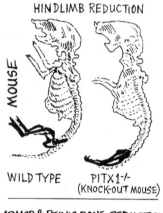

MOUSE

WILD TYPE

PITX1-/-
(KNOCK-OUT MOUSE)

ARMOR & PELVIC BONE REDUCTION

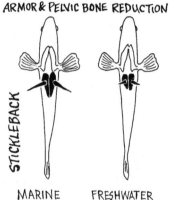

STICKLEBACK

MARINE

FRESHWATER

[Adapted from Michael D. Shapiro et al., "Genetic and Developmental Basis of Evolutionary Pelvic Reduction in Threespine Sticklebacks," *Nature* 428 (2004): 717–723]

Interestingly, laboratory knock out of the Pitx1 gene in mice causes large changes in pelvic structure reminiscent of the armor changes in stickleback. Hence Pitx1 controls bones and appendages in both fish and mice.

DNA & Genetics and the Changing Views of Human Evolution

Modern humans first appeared, according to present-day estimates, about 200,000 years ago. Yet, there is no evidence of any cultural artifacts before 80,000 years ago – and there is telling genetic evidence that human evolution is continuing and that the appearance of the oldest civilizations 5,000 years ago might be associated with genetic changes that we are only beginning to understand.

Recently, scientists studied how rapidly changes occur within the genome and they discovered an area of the genome that has undergone more rapid change than any other area – the Human Accelerated Region 1 (HAR 1). HAR is present in chickens and chimpanzees and only 2 sequence changes have occurred since they separated 310 million years ago. HAR 1 has acquired 18 changes in sequence since humans and chimpanzees separated. (During 5 or 6 million years only one or no changes would occur by chance.) Equally interesting is that the HAR 1 lies outside of the protein coding region – it is a gene-control region. Scientists had long believed that it was the changes in the protein coding regions of the genome that were responsible for the differences between human and chimpanzee brains, but the discovery of HAR suggests that it is changes in the ways genes are switched on and off that is more important in brain development and function than the production of new kinds of proteins. HAR 1 plays a role in embryonic cortical development, the migratory patterns of neurons, and in the adult functioning of the brain.

Human Migration

Patterns of human migration are studied by comparing changes in mitochondrial DNA (which comes exclusively from mothers) and Y-chromosomal DNA (which comes exclusively from fathers) or in the lengths of short DNA repeats (such as CGT repeated 20 times versus CGT repeated 25 times, etc). The greater the differences between these markers in two human populations, the longer ago they migrated as separate groups. Scientists believe that there were a series of migrations of human ancestors from Africa, over the last several million years. One ancestor – *Homo erectus* – appears to have migrated out of Africa some 2 million years ago and settled throughout Europe and Asia. *Homo erectus* had a relatively small brain size [600 to 800 cubic centimeters; chimpanzees: 300-400; humans: 1400]; he may be an ancestor of the Neanderthals. Genetic 'dating' tells us that modern humans all come from a group found in Africa 71,000 to 142,000 years ago and that Eurasian, Oceanian, East Asian, and American populations migrated from Africa some 63,000 or more years ago.

Human Evolution and the DNA Beginnings of Vocal Expression: Birdsong and Language

Not long ago a gene was discovered – FOXP2 (Forkhead box p2) – that encodes a transcription factor and is altered among some members of a family who have marked speech defects. How important the gene is for language function in general is not known. The gene, in a somewhat different form, is found in all mammals. One possible dating of the FOXP2 gene suggests that its present form in humans appeared about 50,000 years ago – about the time of the final migration out of Africa; it might also correspond to the time that human language first developed.

But humans are not the only species with vocal expression. For example, zebra finches learn to sing by imitating the song of another zebra finch, the tutor. At 60-80 days after hatching, the baby finch makes songs that might be compared to a baby's babbling. But by 100 days, the finch's song very closely resembles that of the tutor. The FoxP2 protein, which is encoded by the FoxP2 gene, is expressed in the brains of zebra finches and canaries. FoxP2 protein expression is highest at times in development when the birds are learning to sing as well as at times during the year when they change their songs. Thus, as in humans, FOXP2 gene function seems to be linked to vocal expression. However, not just finches and canaries, but all avian species express the FOXP2 gene, although not all birds sing. And there is no evidence that a bird's ability to sing depends on the evolutionary selection of a special form of

the FOXP2 gene. While FOXP2 gene function is required in some way for vocal expression, its expression certainly is not sufficient. Moreover, the human FoxP2 protein is quite similar to songbird FoxP2, so it would be hard to conclude that the evolution of FoxP2 is the key to humans' unique capacity for language.

Another gene that plays a role in the increase in brain size, ASPM, may have appeared in its present form in humans some 6,000 years ago – just before the appearance of the earliest forms of writing in Mesopotamia. Mutations of the ASPM gene cause the genetic disease microencephaly, in which the brain is reduced to the size of the brain of an early hominid that lived more than a million years ago.

That a mutated gene or set of genes causes a wide range of linguistic problems is not evidence that the mutated gene is a "language gene". A single gene may serve different functions at different times. And language is the outcome of the function of many genes. Like the Hox genes, the FOXP2 gene encodes a transcription factor that regulates other genes and we may have to identify the genes controlled by FOXP2 to understand how FOXP2 contributes to language.

Today, DNA has begun to give us a new way of looking at human history and the origins of human characteristics such as language that distinguish us from other animals. Thus, the study of DNA could be a metaphor for science, knowledge, and understanding. The more we learn about DNA, the more we realize that our notions

about life, genes, inheritance, and evolution are ever changing. When Wilhelm Johanssen first introduced the notion of the gene in 1909, it meant a chemical unit that represented a particular trait – hair color, eye color, etc. – that could be passed on from one generation to the next. By the 1950s, when the structure of DNA became known, it seemed to justify the Darwinian idea of gradualism in evolution; but by the 1980s genes were also (and perhaps most importantly) considered switches – turning on and off other genes and causing changes that did not require gradual accumulations as with the stickleback. And then little bits of RNA were shown to be important in controlling the activities of genes as well.

In a very deep sense, DNA is about our relation to the past, the present, the future and to other living things. And perhaps most of all, DNA is about the puzzle of life: the 'solution' to that puzzle is an on-going, never ending quest. DNA blurs biological time.

In some sense, DNA justifies Albert Einstein's words written four weeks before his death: "The distinction between past, present and future is only a stubbornly persistent illusion."

Epilogue to the First Edition

DNA:
Our Most Valuable Heritage

DNA is the thread which connects us with our most remote ancestors. If there were any interruption in the chain of the inheritance of genetic information, the "evolutionary value" of previous millennia would be lost. The coded genetic information in DNA is the outcome of mutation and environmental selection which together create the evolutionary process. This information could not be replaced.

If we knew every detail of the structure of a cell, apart from DNA, the chemical constitution of the cytoplasm and, nucleoplasm, the lipid content of the membranes, the amino acid sequence of every protein and the folding of the peptide chain, the energy levels of the metabolites and their pathways of degradation, we could not deduce the structure of the DNA. Yet, the DNA of the diploid chromosomes in the undifferentiated egg is sufficient to determine the minute details of the developed organism. By saying that a hen is only an egg's way of making another egg, we emphasize the requirement that as generations pass, the DNA must be transmitted. Germ cell must give rise to germ cell for a species to continue to exist.

A Universal Code

How is our genetic information related to the information of other species? We can clone the genes of many organisms and attempt to rank them on an evolutionary scale. The first fact which is clear is that the genetic code is universal. The genes of virtually all organisms "speak" to the ribosomes and tRNAs of the protein-synthesizing machinery in the same genetic language.

One exception is the code of the mitochrondrion, a cytoplasmic organelle with its own chromosome. Mitochondria are believed to be captured organisms, originally residing symbiotically in the cytoplasm of their host, and now essential components of eukaryotic cells. But even the mitochondrial code is nearly identical with the conventional genetic code. One other possible exception is the *prion*, the protein component of the infectious agent of scrapie, which causes a fatal disease in sheep. The scrapie agent has no evident nucleic acid, hence the speculation that its genetic information is carried in "protein genes" called prions.

These cases aside, the code is universal. There are many variations upon the details of information storage, with viruses providing the greatest deviation. The genome of poliovirus is a single-stranded RNA which serves directly as a messenger RNA. And influenza virus uses helical rods of double-stranded RNA to store its genetic information, with a separate rod for each gene. Provocatively, scientists using cloning technology have made DNA equivalents to poliovirus and influenza genes. When introduced into animal cells, these homologues direct the synthesis of poliovirus and influenza proteins, without any concern that the "stuff" of the original genes was RNA.

What remains constant through these examples is the code itself, which is common to man, bacteria, flu and polio.

Our Ancestry

This universality suggests that all living

things had a common ancestor which bequeathed the code to all of today's living organisms. A gene from a fruit fly need never function in a human cell (although scientists have recently made them do so!). However, flies and men, and their respective ancestors, were endowed with a common genetic code, and too much was at stake (i.e., the expression of their genes) for the codes to change during evolution.

If we dare to speculate about a common ancestor for all living entities, then surely we can rank existing beings, bacteria, fruit flies, man, etc., on some evolutionary scale. Does man, and the animal kingdom, represent the triumph of evolution, when compared, say, with the lowly *E. coli* bacterium?

Bacteria: Relic or High Tech

Conventional wisdom placed the simple bacterium at the bottom of the evolutionary scale. Man has placed himself at the top. The British astronomer Arthur Eddington postulated that "entropy is time's arrow." Using this postulate, we can distinguish the progression of time. For example, if we had a filmstrip showing a building disintegrating, we would know that we were playing the filmstrip forward by observing the building falling apart as time progressed, not reassembling from bits of plaster and timber to form an intact edifice. Is there a similar rule we can apply to determine the direction and the progress of evolution?

Although bacteria seem lower (hence more ancient) than the multicellular eukaryotes, they have some remarkable properties. *E. coli* can reproduce in twenty minutes, about 350,000 times faster than the generation time of man! Bacteria lack the intronic sequences found in eukaryotic genes. They lack the splicing mechanism which removes the intronic sequences during processing mRNA, and they lack the nucleus as well. They possess a single chromosome which is in the cytoplasm. In fact, they lack most features of eukaryotes which make day-to-day life, and the generation of progeny, time consuming. Thus a view which opposes conventional wisdom is that bacteria are highly evolved creatures, "streamlined" for a rapid life style. If bacteria evolved from eukaryotes the streamlining would require that introns be removed from every gene (it would be too much work to evolve the genes from scratch) so that the emerging new prokaryotic life form could dispense with splicing and the nucleus. Indeed, we now know that eukaryotic genes *can* loose their introns by a mechanism in which the mRNA (an "intron-less" form of the gene sequence) is copied back into DNA. Such intron-less genes appear to be fairly common in animal cells, and reverse transcriptase (which we encountered in DNA cloning) or its equivalent could do the job for converting mRNA information (without introns) back into "streamlined" DNA. Evolution of prokaryotes from eukaryotes is not far-fetched!

How Organisms Are Related: The Clue in DNA

If we doubt the ranking of bacterial prokaryotes and eukaryotes in the evolutionary scale, can we ever hope to establish the direction of evolution and thereby glimpse our own origins? Perhaps when the structures of the genes of many organisms have been determined by DNA sequencing (only a million nucleotides are now known) we will be able to perceive a clear structural progression from one species type to another, which will rank these organisms chronologically in the order of their appearance on the evolutionary stage. However, recent evolution of a species and even diversity of biological function (such as playing the cello or reciting Chaucer) are not guarantees of evolutionary fitness. Perhaps after a man-made nuclear holocaust, only cockroaches would survive. A humiliating test of our evolutionary rank!

The Individual and the Species: The Consequences of Mutation

From generation to generation, the adult members of the species die, the differentiated somatic cells are lost, and the informational inheritance in DNA is passed through the germ cells to the next generation.

The species and the individual have somewhat conflicting interests in the transmission of DNA's information, especially in the fidelity with which it is transmitted. The stability of genetic information is in the best interest of the individual. For the individual to survive, he must be well constructed and in good running order. Assuming that the parents of the individual were themselves free from disabling genetic defects, it is in the offspring's interest that he inherit DNA copies of the parents' genomes that are of the highest fidelity. Changes in the nucleotide sequence of the DNA, called mutations, can occur through chemical or X-ray induced damage to the DNA chains. Mutations also arise through base-pairing errors during replication, or through chromosomal rearrangement, that is, transposition of DNA segments

from one chromosomal location to another.

The Individual: "No Errors Please"

To avoid mutational errors, cells employ proofreading functions, built into DNA polymerase, that scan the newly synthesized DNA to edit base-pair errors and correct them. Specific enzymes remove nucleotides damaged by radiation. Thymine dimers, a common example in which neighboring thymine bases in a chain are linked together, are excised and replaced. Despite these precautions, errors occur. Mutations can affect any gene or DNA segment, and alter its expression in unexpected ways. Many changes will have virtually no impact on the individual's ability to survive. A nucleotide change in a DNA region between genes that does not code for protein might have little or no effect on the organism's function. However, nucleotide changes in functional genes will, in general, be deleterious. If a computer manufacturer arbitrarily changed the specifications for the wiring of a circuit board (akin to a random mutation), substituting resistor for capacitor, or transistor for micro-chip, the product most likely would not function. For complex machines, like computers or man, *random* changes will probably be for the worse.

Species
1) Tinkering Permits Variation

The mutability of genetic information is essential to the survival of the species. To continue our analogy, a *very, very* small

236

number of random changes in the design may actually improve the computer. Most computer manufacturers enforce strict quality controls to ensure error-free construction, and employ electrical engineers and market analysts to make carefully planned changes which improve product design. In contrast, nature does not avail itself of analysts and engineers to evolve DNA blueprints. However, the need to evolve remains great. Thus, the species must tolerate tens of thousands of genetic errors imposed upon offspring, causing marginally lower survival value, to obtain, by chance, the one slight genetic improvement that increases survival value.

2) Introns and Evolution

The existence of exons and introns, and the splicing mechanism, provides additional means for evolution. DNA exon segments transposed into introns of genes can add protein coding sequences in mRNA by means of the splicing mechanism. Hence by incorporating new coding sequences, new peptides may be "inserted" into proteins. Should such a new exon encode a functional domain of a protein, this process of "exon shuffling" could provide old proteins with new activities, in a small number of evolutionary steps. Splicing also permits "tinkering" with protein structure at the RNA processing stage of gene expression by varying the coding sequence retained in the mRNA product.

Mutations can arise by many mechanisms. The species as a whole relies on the mutability of its genetic information to permit evolution. However, the individual relies on the invariant, fully faithful

transmission of the same information in germ line cells, to ensure freedom from genetic disability.

3) Cancer and the Genetics of Somatic Cells

Germ line cells provide the DNA for future generations. However, for the individual, genetic transmission also occurs through *somatic* cells. Although somatic mutations have no direct genetic consequences for future generations, our new understanding of cancer's origins show that they can have profound consequences for the individual. Cancer is a somatic genetic disease, an example of natural selection among somatic cells, working against man.

Once we reach adulthood, the number of cells in our body remains relatively constant. Some cells, such as neurons, most of which are "terminally differentiated" and non-dividing, may continue to function for the lifetime of the individual and will not be replaced. Other cells, such as red blood cells, have a finite lifespan. They are routinely inspected in the spleen for imperfections and removed from the bloodstream if found faulty. The "hemopoietic progenitor cells" in the bone marrow retain the ability to divide, and (unlike neurons) produce daughter cells which differentiate to give new red blood cells, thus replenishing the population.

Control of cell division is a crucial feature of the society of cell types which comprise the mature organism. Cells must proliferate only in response to carefully pre-programmed signals. When cell division proceeds unchecked, outside the normal requirements for cell replenishment, a tumor results. If the tumor cell mass disseminates or impairs the function of vital organs, the resulting cancer is life threatening. We now know that many cancers result from somatic mutations which provide cancer cells with an (undesirable) ability to proliferate. Mutations in somatic cells which allow proliferation will, through a cruel form of natural selection, allow the mutant cell type to thrive at the expense of its neighbors and ultimately at the individual's expense too.

Mutations That Cause Cancer

Scientists have used growth selection to identify the genetic changes which convert normal cells to cancer cells. DNA extracted from tumors was introduced into cultured benign cells. Those which took up and expressed the gene for cancerous growth proliferated in the tissue culture dishes of the experimentalists, and are said to be "transformed". This unchecked growth of transformed cells mimics tumor growth in the organism.

The newly acquired gene, called an *oncogene*, which originated in the tumor and which transformed the cultured cells, was isolated by cloning and the protein it encoded was identified. Normal tissues were found to have counterparts to the oncogene, and these were called proto-oncogenes. Proto-oncogenes were also isolated by cloning. Several normal (proto-oncogene) and cancerous (oncogene) gene pairs were identified in this manner. For the best studied example, called the *ras* gene, the DNA nucleotide sequences of proto-oncogene and oncogene were determined and compared. The *ras* oncogene from a colon carcinoma had a single nucleotide change in comparison with its normal proto-oncogene counterpart. A

"G" residue in the codon for the twelfth amino acid of the *ras* gene protein had changed to a "T" residue. While the proto-oncogene specified a protein whose twelfth amino acid was a glycine, the oncogene made a virtually identical protein, but with valine at position twelve.

Why was this nucleotide change important? Both normal and transformed cells make similar *quantities* of the *ras* gene protein. Thus a change in the level of expression is not implicated in cell transformation. Instead, the oncogenic potential of the mutant gene is likely to reside in its specification of an altered protein, with a mutant structure. Investigation of other tumors also revealed cases in which the oncogene was a *ras* gene. In each case, the change in *ras* gene DNA structure, which converted a normal *ras* proto-oncogene to an oncogene, resided in codon twelve. Thus, in other examples, glycine was replaced by aspartate, by serine or by lysine.

We don't fully understand the function of the *ras* gene protein in healthy cells, or why its mutation causes uncontrolled proliferation. Apparently the *ras* protein plays a crucial role in cell growth control. When glycine, a small amino acid, is replaced by any of several bulky ones, the *ras* protein function may be impaired. Most likely the mutant protein chain cannot fold to assume a fully functional structure. Like a minute engineering error which causes an automobile brake to fail, the small change caused by the *ras* oncogene mutation can lead to disaster!

The recent discoveries of oncogenes follow directly from the technological advances in genetic engineering and DNA manipulation. Through implementation of gene transfer, tissue culture selection, DNA cloning and sequencing, we can lay bare the structure of any gene or its mutants, and predict the sequence of its protein product. Ironically, we can generate this new information more rapidly than we can assimilate and interpret it. Thus, we know that the mutant *ras* oncogene is a "cancer criminal". We are still unsure of the exact nature of the crime against cell growth regulation that the mutant *ras* has committed.

Are We More Than What DNA Has Made Us?

Since it is now possible, in principle, to decipher the genetic structure of any organism, we can ask ourselves to what extent is the organism determined directly by the DNA? For example, is DNA complexity directly related to that of the organism? While earlier studies on bacterial genes gave the impression that most bacterial DNA codes for protein, scientists were quite surprised and even puzzled by the discovery that for animal cells, only 10 percent of the DNA is protein coding. In fact, the function of most of the DNA in animal cells remains unknown, and some scientists have suggested that it doesn't have any functional role at all. Whatever the case, we can determine the total quantity of DNA in cells from different organisms and relate this to morphological complexity. We find that a frog cell has 3.5 picograms of DNA, a human cell 3.4 picograms, and that of a lily flower 32.8 picograms. From these figures, we see that DNA content is not a simple index of the structural complexity of an organism. Furthermore, a mouse brain has some six million cells, whereas a human brain has tens of billions of cells. There is no hint of this in the DNA content of these cells.

We have already asserted that the DNA of an organism, transmitted through the germ cells, determines the final morphology of the adult. If we knew all of the structural details of the genes of an organism, including the complete nucleotide sequence of the DNA, would we be able to predict this morphology? How in fact does DNA act to determine the structure of the organism?

We know that DNA determines the linear sequence of amino acids in the protein translation product. Yet, the essential characteristics of proteins depend not just upon their amino acid sequences, but also upon the detailed three-dimensional structure of the folded protein chains. Subtle physical and chemical interactions between different amino acids of the chain, and between the chain and its chemical environment, determine the precise nature of the folding. Although we can deduce the linear primary sequence of amino acids from the DNA, to calculate the way the protein folds is at the limit of our current capabilities. Second, if we knew nothing about the gene save its structure, we might have great difficulty deducing the function of the protein product, for example, its catalytic activity, were it an enzyme.

If we consider that any cell has tens of thousands of different proteins, and that the body is composed of tens of billions of different cells, the computational problem becomes immense. Finally, to deduce the various metabolic reactions and activities which constitute the normal routines of the cell would be immeasurably beyond our capabilities. Thus, despite the great power of the "new biology" and its techniques, it alone is insufficient to unravel the molecular mechanisms of development.

Fortunately, more traditional genetics provides us with some insights. A simple way to study development is to analyze organisms with mutations in specific genes that control differentiation, and to note the consequences in the adult organism. We will find that what is altered is more complex than we would have predicted from simple bacterial models. Nature has already done such experiments for us, and one example is the condition called *albinism*.

Genetic Case Studies
1) Albinism

Among the striking features of this condition are light skin and red eyes in humans, or the distinctive pattern of light and dark patches in the Siamese cat. The site of the mutation is in a gene believed to determine the structure of an enzyme, *tyrosinase*. Tyrosinase plays a crucial role in the synthesis of the pigment melanin. The light color of the skin (or fur, in cats) is a direct consequence of the absense of melanin. But that is only part of the story. Melanin also is found in the epithelial layer of the eyes. Albinos lack the pigment and have red eyes. For reasons that are poorly understood, the failure of melanin to appear in the epithelial layer behind the retina causes the optic nerve (which normally partly crosses as it projects to the brain) to project in an abnormal manner. (This is seen in Siamese cats as well as humans.) Albinos have either very restricted visual fields and/or only monocular vision. What is apparently a simple mutation affecting skin pigmentation causes a cascading effect which includes severe neurological problems. A single gene has consequences for many aspects of an organism's development. In keeping with this complexity, the gene that codes for the enzyme tyrosinase cannot be said to code directly for vision or, for that matter, for the circuitry of the brain.

2) Genes Versus Environment in Sexuality

Animal behavior also depends on both genetic and environmental factors. For example, the development of animal and human sexuality is not as clearly "genetically determined" as one might believe. The hormonal environment of the developing fetus can have profound effects on later sexual development and activity.

The characteristic position of the female rat during sexual intercourse is known as *lordosis*. The female arches her back and sweeps her tail aside in order to receive the male sex organ. Males, on the other hand, exhibit mounting behavior. Male rats that are castrated at birth will exhibit lordosis as adults rather than mounting behavior. Therefore, a rat that is genetically male will behave as a female. This can be prevented by giving the rat injections of male hormones during the period immediately following its birth. If the hormones are given later in life, they will fail to suppress the lordotic posture in intercourse. In the female, lordosis is eliminated when the ovaries are removed.

These crude experiments only suggest the subtle variations in sexual behavior that result from differing hormonal environments of the fetus and later in development. Female and male brains have been found to be anatomically and neurochemically different and these differences develop during the *in utero* period. Female brains are modified when male hormone is present in the uterus because of a male litter mate.

Genetic Programs for Behavior

While environmental factors, be they *in utero* or post-natal, certainly profoundly affect the manner in which genes are expressed, rather complex, apparently genetically determined patterns of behavior have been known for some time.

The Dutch ethologist Nikolas Tinbergen, now living in England, showed many years ago that some animals engaged in certain relatively fixed patterns of behavior. In one famous example, the stickleback fish, Tinbergen demonstrated that the mating ritual followed a fixed sequential set of acts, each aspect of which was initiated by a definite physical signal. The male stickleback, for example, acquired a red spot on its belly during the mating season. The female will follow a male with this sign to its nest. Any more or less fish-shaped object with a red spot could lead the female on. If the object is turned upside-down, so that the spot now appeared on the top, the female fails to follow. The entire mating procedure in the stickleback required clear signs (such as the red spot on the belly of the male) presented in the proper order. Whether these signs are produced by the mating fish or a mechanical object does not matter. The appearance of the sign produces the appropriate behavior. The implication of Tinbergen's work was that this behavior is not learned but rather is genetically controlled.

Recent studies on *Aplysia*, a shell-less marine snail, which grows up to five inches long, have in fact isolated some of the genes that control mating

behavior in that animal. Egg-laying behavior consists of a specific series of actions, a rigid behavioral pattern which is phenotypic of the species: copulation, ejecting an eggstream from a duct, increasing heart beat, waving the head, catching the eggstream in its mouth, winding the stream into a solenoid, and placing it on a rock. Peptide hormones, several amino acids long, produced in the "bag cells" of *Aplysia* will induce this behavior if injected into a virgin animal. The laboratories of Richard Axel and Eric Kandel cloned the gene for the egg-laying hormone. From its DNA sequence they deduced that the peptide hormones responsible for this behavior are produced by cleaving a larger protein which is synthesized in the bag cell. Their work suggests that different peptide hormones cleaved from the larger peptide activate different steps in the egg-laying ceremony. Thus a gene directly determines the onset of these actions, with different segments of the gene apparently responsible for different components of the activity.

Behavior Beyond the Reach of Genes

These examples of genetic determinism far from prove that all behavior is genetically controlled. Genes do not operate in a vacuum. The stickleback's rituals are initiated by different environmental cues. If they fail to appear, so will the ritualized acts. Of course, in principle, we could alter the nature of the performance by altering one or two genes, as might be possible with *Aplysia*. But most behavior is not the consequence of a specific gene. Our linguistic ability certainly has a genetic basis in the organization of the brain, but the language that we speak is not determined by our genes. Children of English-speaking parents are not born with "English language" genes anymore than children of Japanese parents are born with "Japanese language" genes. Rather we are born with a capacity to learn *any* language to which we might be exposed. There are no genes for a specific langauge, and likewise there are no genes for specific thoughts either. Language and thought — though dependent on a genetic *capacity* to learn — are also the consequences of environmental factors which cannot be programmed genetically.

Bioengineering Cures for Genetic Disease

There are specific mental disorders (such as schizophrenia) that may be the consequence of physiological dysfunction. Such diseases are today controlled through medication, with varying results. Part of the problem is that we are not as yet sure just what the physiological malfunction is or whether it has a genetic basis. But in the cases that diseases have a clear genetic cause, would engineering be of any use in curing the disease? Sickle cell anemia is an example of a disease in which nucleotide changes in the beta globin gene yield a mutant globin protein, with altered oxygen-binding properties. Does the new technology provide us with any procedures to correct malfunctioning genes?

We are still at the early experimental stages of clinical applications, but an experiment indicates some of the future possibilities. In late 1982, a team of scientists reported that they had produced mice that were almost twice normal size.

Growth hormone is a protein that has profound effects on the development of cells, and among the more deleterious effects of abnormally high levels is the production of giantism in humans. The scientists therefore decided to see if they could introduce the gene for growth hormone into a mouse embryo, activate the gene, and produce an abnormally large mouse. Since the control regions of the growth hormone might have prevented its activation in most mouse tissues, the gene was attached to the transcription control region of the mouse metallothionein gene — a gene that is turned on by the presence of toxic heavy metals. The advantage of this promoter control region is that it is active in most tissues of the mouse, though the level of expression varies from tissue to tissue. The assumption, then, was that the metallothionein gene promoter would be active in most of the cells of the body of the mouse and that it would in turn activate the growth hormone gene to which it was attached.

The scientists injected the metallothionein promoter linked to the growth hormone gene into the pronucleus of a fertilized mouse egg and then inserted the egg into the reproductive tract of a foster mother. The foreign DNA fragment was apparently integrated into the embryo genome, and because its promoter was from the metallothionein gene, the growth hormone gene was effectively expressed, resulting in a "giant"mouse.

The success of this experiment suggests the possibility of new therapeutic procedures for correcting genetic defects as well as some perhaps more questionable ones of creating giant cows, fowl, etc. But it also points to the limitations of our present technology.

Growth hormone was expressed throughout the body of the giant mice. Many genetic defects are specific to particular cell types and at present there is no way of turning genes on within a given category of cells. A way might yet be found, but then present techniques require that the "corrected" gene be introduced into the embryo. This would appear to limit its clinical usefulness, since we are unlikely to know enough about the genome of an unborn child to want to manipulate its genes while it is still an embryo at the earliest stages of development. The technique may eventually prove useful in the treatment of general somatic disorders of genetic origin, though there are many problems yet to be solved.

There is no way of knowing how future discoveries will modify our present understanding of the nature of DNA. Science is a bag of surprises. And it is the surprises that maintain our endless curiosity.

However, DNA is only part of the story. It contains the code for the linear sequence of amino acids that make up

protein structure. But the clues it provides about the morphological and chemical characteristics of the organism are obscure. The fossil bones of dinosaurs tell us of the morphology of that long extinct species. Should one day a preserved piece of DNA be found among those fossilized remains, given our present knowledge and technology, we could hardly re-create the beasts. Indeed, if nature had left us the DNA and not the fossils, we would not have been able to imagine a dinosaur and we would have had no clue about its size. At best, we might have been able to clone a dinosaur globin gene and compare it to other modern globin genes.

Through DNA we have revolutionized our understanding of life. We have opened the door to new technologies that may be of great benefit medically, and even industrially. However, DNA on its own is but the plan without a context. Complete organisms are far more complex than DNA. The discovery of DNA has excited our curiosity and stimulated our desire to know more. An incredible molecule, for sure, it continues to raise as many questions as it has answered.

GLOSSARY

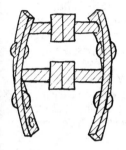
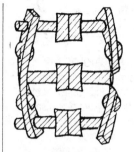
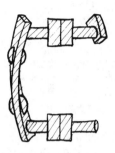

AMINO ACID
The fundamental building block of proteins. There are twenty different amino acids (for example, glycine, alanine, and lysine), which are linked during protein synthesis on the ribosome according to the coded genetic information in messenger RNA. The link that joins one amino acid to the next in the protein chain is called a peptide bond.

ANTICODON A triplet of bases in transfer RNA that pairs with a codon in messenger RNA. *See* **codon**.

ARM (OF CHROMOSOME) A segment of a chromosome that projects from the centromere, either the long arm or the short arm.

BACTERIOPHAGE (PHAGE) A virus that infects bacteria. Bacteriophage consist of a double- or single-stranded DNA or RNA genome wrapped in a protective protein coat. The name comes from the Greek word *phagos*, which means "one that eats."

BIOTECHNOLOGY The use of genetic engineering and other new biology techniques for commercial purposes.

CENTROMERE DNA region where sister chromatids make contact.

CHROMATIN The complex of DNA with protein that resides in the living cell. In eukaryotes, the fundamental structural unit of chromatin is the nucleosome. *See* **nucleosome**.

CHROMOSOME A large chromatin structure that consists of a highly folded DNA chain, complexed with basic proteins. In eukaryotes, chromatin condenses during mitosis into distinct X-shaped structures that independently segregate, as the cell divides, between the daughter cells.

CLONE A population of cells that arises from a single mother cell and thus are genetically identical to one another.

245

CODON A triplet of bases in messenger RNA that codes for an amino acid. *See* **anticodon**.

COMPLEMENTARITY The relationship of the DNA sequence of one strand of a double helix to the sequence of the other strand. When the base guanine in one strand faces cytosine in the other, and adenine faces thymine, as dictated by the base-pairing rules, the two chains are said to have complementary sequences.

CONSTITUTIVE MUTANT A class of mutants of a regulated gene that synthesizes the gene product, whether or not the inducer is present. *See* **inducer** *and* **repressor**.

CORE PROCESSES Genetic, biochemical, or other processes that are conserved through evolution, making the rapid development of new organism possible.

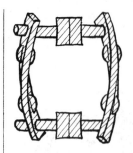

DEEP HOMOLOGY A similarity that underlies the genetic control of body pattern in two organisms that is not evident on morphological or phylogenic grounds.

DIPLOID As applied to a cell, possessing two complete sets of chromosomes. *See* **haploid**.

DNA HELIX: A, B, and Z FORMS DNA can assume different double-helical structures, depending on the solvent conditions and the nucleotide sequence. These structures were originally seen when DNA fibers were analyzed under different states of hydration by X-ray diffraction. In the A form, which is favored at low humidity, the helix is right-handed, but the plane of the bases is inclined with respect to the axis of the helix. At higher humidity, and most likely in the living cell as well, the prevailing structure is the B form, also a right-handed helix, but with the plane of the bases nearly perpendicular to the helix's axis. When the DNA sequence alternates

between purines and pyrimidines (such as GCGCGCG . . .), a left-handed helix called the Z form is stable.

DNA METHYLATION The addition of a methyl group (CH_3) to a DNA base, often to cytosine.

DOMAIN A structural segment of a protein molecule that is created when the peptide chain folds.

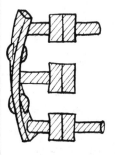

ENZYME A protein molecule that catalyzes biochemical reactions. Examples are beta-galactosidase, which catalyzes the hydrolysis (cleavage with the addition of water) of specific bonds in sugars called beta-galactosides, and RNA polymerase, which catalyzes the linkage of ribonucleotides to one another to make an RNA chain. Enzymes differ from synthetic catalysts in that they exhibit exquisite specificity in the reactions that they catalyze and that they function under physiological conditions.

EPIGENESIS The doctrine that the development of the body is determined by the interaction of the genes with the environment.

EPIGENETICS Heritable changes in an organism that do not involve changes in the sequences of DNA. In epigenetic mechanisms, environmental factors may alter DNA base structure (such as methylation of cytosine), and hence DNA function, but the base sequence itself is not changed..

EUKARYOTES Organisms – including plants, animals, protozoa, and fungi – whose cells have nuclei. *See* **prokaryotes**.

EXON A continuous segment of a eukaryotic gene whose sequence is retained in mRNA and that usually encodes protein. Many eukaryotic genes are "split" and have exons interspersed with nonsense DNA called introns. *See* **intron**.

EXPLORATORY BEHAVIOR A random behavior that, when useful features are selected, can give rise to a precise behavioral pattern.

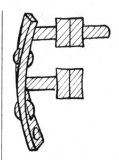

FERTILIZATION The fusion of the sperm and the egg.

F-FACTOR (FERTILITY FACTOR) A piece of DNA that confers "maleness" on a bacterium.

FRAME-SHIFT The deletion or insertion of one or more bases in the coding region of a gene that causes incorrect triplets of bases to be read as codons.

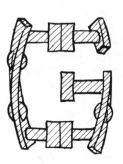

GENE A sequence of DNA that codes for a functional product. Most genes encode proteins, but genes also encode such RNAs as transfer RNA. Genes are the basic units of heredity.

GENETIC CODE The code used by living organisms to store genetic information, by which triplets of bases in DNA (or messenger RNA) represent amino acids in proteins.

GENOME The total genetic information of a cell or virus as represented by the DNA. Some viruses have RNA genomes.

GENOTYPE The genetic constitution of an organism. *See* **phenotype**.

GERM CELLS Cells that give rise to the gametes (sperm and eggs) and thus transmit genetic information to succeeding generations. They are formed early during the development of the embryo and eventually divide through meiosis to yield the gametes.

GRADUALISM The Darwinian view that evolution takes place though the accumulation of small changes over extremely long periods of time.

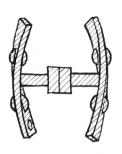

HAPLOID The genetic content of one set of chromosomes. Sex cells (gametes) are haploid, and in some organisms (bees and wasps) somatic cells are also haploid. Upon fertilization, the haploid egg receives a second set of chromosomes from the haploid sperm, producing a diploid cell. *See* **diploid**.

HIGH THROUGHPUT SEQUENCING The modern method of DNA sequencing that can analyze millions of nucleotides in a single session and that is often massively parallel – that is, with many short DNA sequences analyzed simultaneously.

HISTONES Proteins rich in basic (that is, positively charged) amino acids that are found in the chromosomes. There are five fundamental histone types. Nucleosomes consist of a helix of DNA wound around a core of histones.

HOMEOTIC MUTATION Mutation, often in a Hox gene, in which the body plan of an animal is reorganized and, for example, one body part is replaced by another.

HORMONES Substances (often small polypeptides or proteins) that are synthesized in one group of cells in the body and then are released into the body to affect the functioning of other cell types (or organs) in the body.

HOX GENES A family of genes that controls the development of the body plan in all bilateral animals.

HYBRIDIZATION The formation of double-stranded DNA–DNA or RNA–RNA complexes from a mixture of single-stranded DNA or RNA. Hybrids are formed only if the base sequences of the DNA or RNA strands are complementary. Hybridization also is a technique for determining the similarity between the base sequences of two nucleic acid molecules.

IMMUNOGLOBULINS Antibodies that consist of "light" and "heavy" protein chains bound in a Y-shaped structure.

IMPRINTING An epigenetic mechanism in which the expression of a gene depends on whether it was inherited from the mother or the father.

INDUCERS Small molecules (often metabolites such as sugars) that bind to a repressor, releasing it from an operator. *See* **operator** *and* **repressor**.

INSULIN A polypeptide hormone secreted by specialized cells in the pancreas that regulates metabolism and the production of energy. It stimulates the uptake of glucose by muscle cells and the synthesis of protein. Insulin was the first protein whose amino acid sequence was determined, a feat accomplished by Frederick Sanger.

INTELLIGENT DESIGN The dogma that an intelligent forces contributed to the origin of life.

INTRON A DNA sequence in eukaryotes that lies within genes, but does not code for proteins. Most introns have no apparent function. Their presence "splits" the coding region of a gene into segments called exons. In the synthesis of messenger RNA, introns are copied into RNA, but they are removed by splicing, which restores the continuity of coding sequences. *See* **exon**.

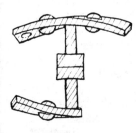

JUNK DNA DNA without apparent function. Approximately 90% of animal cell DNA does not code for protein and is speculated to fall into this class. *See* **intron** *and* **selfish DNA**.

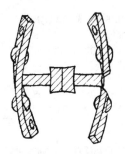

KNOCK-OUT MOUSE A mouse from whose genome a specific gene has been permanently removed.

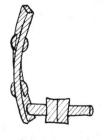

LAMARKISM The doctrine, held by the French naturalist Jean-Baptiste Lamarck, that acquired characteristics are inherited and may result from mutations that are responses to an organism's environment. Most biologists consider Lamarkism to be wrong; however, this view has been reevaluated in special instances of epigenetic inheritance. *See* **epigenetics**.

LIGASE An enzyme that links DNA molecules by connecting an end of one linear double-stranded molecule to an end of another linear double-stranded molecule to create a continuous double helix.

LIPIDS Water-insoluble molecules – such as steroids, fatty acids, and waxes – that are important components of cell membranes and store energy.

MEIOSIS The process by which a cell gives rise to daughter cells with half the number of chromosomes as that in the mother cell. (Diploid cells become haploid.) The sex cells (gametes) are produced through meiosis.

MESSENGER RNA (mRNA) An RNA molecule that is transcribed from a gene and that contains the coded information for the amino acid sequence of a protein. The information in mRNA is translated on the ribosome. In prokaryotes, one mRNA can code for more than one protein.

MICRO RNA (miRNA)
A short RNA molecule (21 to 23 nucleotides long) that can regulate the expression of genes.

MITOSIS
The stage in the life cycle of a eukaryotic cell during which sets of chromosomes destined for daughter cells separate and cell division takes place.

MODERN SYNTHESIS
The synthesis of Darwinian theory and genetics in which evolution is the consequence of the accumulation of small random changes in genes.

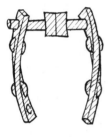

NATURAL SELECTION
In evolutionary theory, the process by which the adaptation of a population to its environment is improved. A large number of variant forms are produced (through DNA recombination, sexual reproduction, mutation, and so on) and the organisms best adapted to their environment survive and reproduce, passing on their genetic material.

NONSENSE MUTATION
A mutation that changes a codon into a three-base sequence that does not specify an amino acid. Such triplets, known as nonsense codons, are UGA, UAA, and UAG.

NUCLEOSOME
The repeating structural unit of chromatin, consisting of 200 base-pairs of DNA wrapped around a histone core. The nucleosomes, plus the DNA that links them to one another, make up the chromatin fibers of chromosomes. *See* **chromatin** *and* **chromosome**.

NUCLEOTIDE
The fundamental unit of the DNA (or RNA) chain. Nucleotides consist of the base (adenine, guanine, cytosine, or thymine in DNA, with thymine replaced by uracil in RNA) plus the sugar (deoxyribose in DNA, ribose in RNA) and linked phosphate.

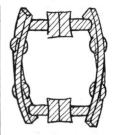

ONCOGENE
A gene responsible for the transformation of normal cells into cancer cells. Almost all oncogenes are mutant versions of cellular genes.

OPERATOR
In bacteria, the site on DNA to which the repressor binds, preventing RNA polymerase from transcribing the operon. Thus it is a control site for transcription. *See* **operon** *and* **repressor**.

OPERON
The group of adjacent structural genes controlled by an operator and a repressor. *See* **operator** *and* **repressor**.

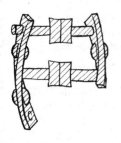

PEPTIDE A chain of amino acids.

PEPTIDE BOND A link (covalent bond) between two amino acids in protein.

PHENOTYPE
The characteristics of an organism as manifested in its developed form. The phenotype is the consequence of the interaction of genes with the environment. *See* **genotype**.

PICOGRAM One-millionth of one-millionth of a gram.

PLASMID A small, circular DNA molecule (typically about 500 nucleotides long) that replicates in a bacterium independently of the bacterial chromosome.

POLYMERASE
An enzyme that catalyzes the polymerization of nucleotides into long nucleic acid chains. RNA polymerase synthesizes RNA; DNA polymerase, DNA.

POLYMERASE CHAIN REACTION (PCR)
An enzymatic technique for replicating specific DNA sequences in a test tube.

PRIMER A short DNA or RNA chain, base-paired to a complementary DNA strand, that is elongated by DNA polymerase. The 3' terminus of the primer is the acceptor for the newly added nucleotide residues and is the starting point for DNA synthesis. Reverse transcriptase also uses an RNA primer, but employs RNA as the template.

PRION A protein that exists in both a nontoxic form and a misfolded, toxic form. The misfolded form can induce the refolding of the nontoxic form. When a healthy animal consumes the toxic form of a prion, its own prion protein can become misfolded and toxic.

PROBE A single-stranded DNA or RNA molecule with a specific base sequence (usually between 100 and 1000 nucleotides long) and tagged with a marker, such as radioactivity or a fluorescent molecule. Hybridization of the probe through base-pairing can detect an RNA or a DNA molecule with the complementary sequence.

PROKARYOTES Organisms whose cells lack nuclei, including bacteria and cyanobacteria (blue-green algae). *See* **eukaryotes**.

PROMOTER The site on DNA where RNA polymerase binds and initiates transcription. More properly, it is defined genetically as a site whose mutation alters the rate of transcription of an adjacent gene.

PROTEIN A biological molecule consisting of amino acids linked into chains. Proteins may have more than one chain, range from tens to thousands of amino acids in length, and serve the cell as enzymatic catalysts or structural components.

PSEUDO-GENE A gene that is nonfunctional, most often as a result of mutational damage incurred during evolution. Pseudo-genes may arise through the duplication of functional genes, followed by the divergence of one copy through mutation such that it no longer may be expressed. Some pseudo-genes are formed by copying messenger RNA into DNA and then inserting the copy back into the chromosomes. These lack introns, have the spliced structure of mRNA, and are called intron-less pseudo-genes.

PURINES Organic bases that contain both carbon and nitrogen atoms arranged in a two-ring structure. Adenine and guanine are purines found in DNA and RNA, linked respectively to deoxyribose and ribose.

PYRIMIDINES Organic bases that contain both carbon and nitrogen atoms arranged in a single-ring structure. Thymine and cytosine are pyrimidines found in DNA, linked to deoxyribose. In RNA, the pryimidine uracil replaces thymine and, like cytosine, is linked to ribose.

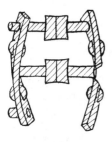

RECOMBINATION The rearrangement of DNA such that sequences originally present on two DNA molecules are found on the same molecule. With homologous recombination, the transfer is between two very similar (but not necessarily identical) DNA molecules. With heterologous recombination, the transfer is between DNA molecules

unrelated in nucleotide sequence. Recombination may take place by the breakage and reunion of DNA molecules.

REGULATORY GENE A gene that encodes a protein or another factor that regulates the activity of a second gene.

REPLICATION FORK The position on DNA where replication takes place. The parental DNA strands diverge at the replication fork to serve as templates for daughter DNA synthesis, creating a Y-shaped form.

REPRESSOR A protein encoded by a regulatory gene that can either combine with an inducer, permitting the transcription of structural genes, or bind to the operator blocking access of RNA polymerase to the promoter, thereby repressing transcription. *See* **inducer**, **operator**, *and* **promoter**.

RESTRICTION ENZYME An enzyme that cleaves DNA at short, specific sequences. Examples are EcoR1 (*E. coli* restriction enzyme 1) and HindIII (isolated from the microorganism *Haemophilus influenzae*, serotype d), which cut, respectively, at GAATTC and AAGCTT in double-stranded DNA. The very high sequence specificity

for restriction enzyme cleavage makes these enzymes excellent tools for dissecting DNA.

REVERSE TRANSCRIPTASE An enzyme in certain animal viruses called retroviruses. Starting at an RNA primer, reverse transcriptase will make a DNA copy of an RNA template, a process that is crucial to the retrovirus life cycle and useful to the genetic engineer in making DNA clones of messenger RNA. The flow of information from RNA to DNA is the reverse of the normal information pathway, hence the names *reverse* transcriptase and *retrovirus*.

RIBOSOME The microparticle in the cytoplasm that consists of RNA and protein, where messenger RNA is translated into protein.

RNA INTERFERENCE (RNAi) The regulation of gene expression, most often through gene repression, by micro RNAs or genetically engineered short hairpin RNAs (shRNAs), which function similarly to the natural miRNAs.

SELFISH DNA Genes proposed to proliferate within the genome to many hundreds of thousands of copies, but do not serve a function for the organism. Selfish genes are thus parasites of the genome and represent the ultimate self-centered biological substance. *See* **junk DNA**.

SEXUAL CONJUGATION The transfer of the bacterial DNA chromosome from a male to a female bacterium. Male bacteria contain plasmids called F-factors, which mobilize this transfer. *See* **F-factor**.

SINGLE NUCLEOTIDE POLYMORPHISMS (SNPs) Single nucleotide variations in DNA structure that appear with greater than 1% frequency and can be used as genetic markers – for example, as markers for disease genes.

SINGLE-STRANDED DNA (or RNA) A DNA (or an RNA) chain whose bases are not paired with those on a complementary chain. Unlike double-stranded nucleic acid, which forms relatively rigid, elongated structures, single strands are floppy and can easily coil back on themselves.

SOMATIC CELL In multicellular organisms, a cell of the soma, or tissues, as opposed to a cell of the germ line. Somatic cells divide and differentiate during development, but under normal circumstances do not exchange genetic information.

STEM CELL An undifferentiated embryonic cell that has the capacity for self-renewal and that gives rise to any specialized cell type. Totipotent stem cells can give rise to all cell types of the embryo as well as cells of the extra embryonic tissue, while pluripotent stem cells, such as embryonic stem cells (ES cells), can give rise to only a subset of these cell types, such as those of the embryo itself.

STRUCTURAL GENE A gene that codes for a protein, such as an enzyme, or for an RNA, such as transfer RNA or micro RNA.

SV40 (Simian Virus 40) A virus of monkeys that also infects tissue-cultured cells in the laboratory and has served as a model for gene expression in animal cells. SV40 DNA is a double-stranded circle with approximately 5200 base-pairs.

TELOMERE A region at the end of a chromosome made up of short, repeated DNA sequences that protects the chromosome.

TETRANUCLEOTIDE HYPOTHESIS A hypothetical structure for the DNA molecule proposed by the organic chemist P. A. T. Levene, in which the DNA nucleotides adenine, guanine, thymine, and cytosine are arranged in a monotonous repetition of short simple sequences (such as AGTCAGTCAGTC . . .). This hypothesis was incompatible with an informational role for DNA.

THYMINE DIMER
Two thymine bases on adjacent nucleotides of a DNA strand may be joined by covalent bonds through the action of ultraviolet light or X-rays to form a thymine dimer. If the dimer is not excised and the DNA is not correctly repaired, a mutational change in the DNA sequence will appear at the site of the dimer.

TRANSCRIPTION
The synthesis of RNA by RNA polymerase, directed by the template strand of DNA. It is a fundamental step in the utilization of genetic information.

TRANSCRIPTION FACTOR
A protein that binds to DNA and regulates the expression of a neighboring gene.

TRANSFER RNA (tRNA)
A class of small RNA molecules that function in protein synthesis. Transfer RNA interprets the genetic code information in messenger RNA.

TRANSFORMATION
As used by Oswald Avery, the term "transformation" refers to the transfer of genes in the form of chemically pure DNA to a cell such that they are integrated into the cell's genome and are functionally expressed. This is also known as transfection. Transformation also refers to the changes in an animal cell's growth properties and morphology that occur when the cell changes from a healthy into a cancerous cell. This can occur when the cell acquires an oncogene. Cells with unchecked growth are said to be transformed.

TRANSGENIC MOUSE
A mouse into whose genome a foreign DNA sequence, often a gene, has been added.

TRANSLATION
The reading of the genetic code in messenger RNA during the synthesis of protein. This is performed by transfer RNA, on the ribosome, and leads to the assembly of amino acids into a chain, with their sequence dictated by the DNA sequence of the gene.

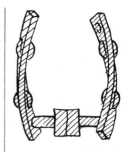

VARIABLE NUMBER OF TANDEM REPEAT SEQUENCES
(VNTRs) Short DNA sequences (9 to 80 nucleotides long) that are repeated frequently in a way that is unique to each individual and whose analysis can identify the person who is the source of a DNA sample.

VECTOR
A DNA molecule or a virus that can be employed to introduce a foreign gene into a cell. Transfers to bacterial or eukaryotic cells use different types of vectors. Vectors can be designed and constructed by a molecular biologist using genetic-engineering techniques. Most DNA vectors for use with bacteria are derived from bacterial plasmids, Lambda phage, or other genomes that replicate within the cell. They often carry genetic markers, such as a gene conferring resistance to an antibiotic. Transfer to a eukaryotic cell may employ simple DNA vectors or viral vectors, such as sindbis virus, lentiviruses,

or adenovirus. Viral vectors contain an RNA or a DNA genome wrapped in a protein coat. In vector form, these viruses have been engineered to accept the foreign gene chosen for expression in the cell. Most viral vectors lack one or more of the virus's own critical genes and thus are crippled and cannot carry out their normal replication cycles.

VIRUS A simple microscopic organism that consists of genetic information (most often DNA, but sometimes RNA) wrapped in a protein coat. In order to replicate, viruses must enter a host cell (animal, plant, or bacterium, depending on the virus type) and divert the gene expression machinery (that is, the ribosomes, transfer RNA, and so on) to the manufacture of new viruses. Viruses that replicate in bacteria are often called bacteriophage (phage).

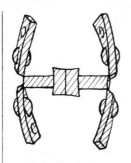

X-RAY DIFFRACTION
When a beam of X-rays passes through a crystal, the beam interacts with the regular array of atoms in the crystal. It exits from the crystal as a complex group of beams. The intensities, angles of exit, and phases of these beams are directly determined by the atomic structure of the repeating unit of the crystal. When changes imparted by the crystal to the beam are determined, the structure of the repeating unit of the crystal may be calculated. This method of structural analysis, called X-ray diffraction, was developed at the Cavendish Laboratory at Cambridge University by William Henry Bragg and his son William Lawrence Bragg. The structures of many proteins, small RNAs (such as transfer RNA), and the DNA double helix were determined by X-ray diffraction.

Reading List

Carroll, Sean B. *Endless Forms Most Beautiful: The New Science of Evo Devo and the Making of the Animal Kingdom.* New York: Norton, 2005.

Carroll, Sean B., Jennifer K. Grenier, and Scott D. Weatherbee. *From DNA to Diversity: Molecular Genetics and the Evolution of Animal Design.* Malden, Mass.: Blackwell, 2005.

Jablonka, Eva, and Marion J. Lamb, with illustrations by Anna Zeligowski. *Evolution in Four Dimensions: Genetic, Epigenetic, Behavioral, and Symbolic Variation in the History of Life.* Cambridge, Mass.: MIT Press, 2005.

Judson, Horace Freeland. *The Eighth Day of Creation: Makers of the Revolution in Biology.* Expanded ed. Plainview, N.Y.: Cold Spring Harbor Laboratory Press, 1996.

Keller, Evelyn Fox. *A Feeling for the Organism: The Life and Work of Barbara McClintock.* San Francisco: Freeman, 1983.

Kirschner, Marc W., and John C. Gerhart. *The Plausibility of Life: Resolving Darwin's Dilemma.* New Haven, Conn.: Yale University Press, 2005.

Maddox, Brenda. *Rosalind Franklin: The Dark Lady of DNA.* New York: HarperCollins, 2002.

Watson, James D. *The Double Helix: A Personal Account of the Discovery of the Structure of DNA.* New York: New American Library, 1969.

ISRAEL ROSENFIELD received an M.D. from the New York University School of Medicine and a Ph.D. from Princeton University. He is a professor at the City University of New York and his books, which have been translated into a number of languages, include *The Invention of Memory: A New View of the Brain*; *The Strange, Familiar, and Forgotten: An Anatomy of Consciousness* (revised and expanded French edition, 2005); and the satirical novel *Freud's "Megalomania,"* a *New York Times* notable book of the year. He has been a Guggenheim Fellow and a longtime contributor to the *New York Review of Books*.

EDWARD ZIFF studied chemistry at Columbia University and received his Ph.D. in biochemistry at Princeton University, where he met Rosenfield. He then joined the laboratory of DNA-sequencing pioneer Fred Sanger in Cambridge, where Ziff helped to develop the first DNA-sequencing techniques. Ziff has been a Howard Hughes Medical Institute investigator, and his research includes many "firsts" in the areas of gene structure and control, cancer biology, and, more recently, brain function. He has written for the *New York Review of Books* with Rosenfield about evolution and the brain. He is professor of biochemistry and neural science at the New York University School of Medicine.

BORIN VAN LOON has been a leading freelance illustrator since 1977. He has designed and illustrated fifteen documentary comic books on subjects from Darwin to psychotherapy and from the Buddha to statistics. Collagist and surrealist painter, he has worked for a wide variety of clients in editorial, publishing, and promotion. He created an eclectic collage/cartoon mural on the subject of DNA and genetics for the Health Matters Gallery in London's Science Museum.